Industrial Engineering

Industrial Engineering

Geoffrey Williamson

Willford Press
www.willfordpress.com

Published by Willford Press,
118-35 Queens Blvd., Suite 400,
Forest Hills, NY 11375, USA

ISBN: 978-1-64728-333-9

Cataloging-in-Publication Data

Industrial engineering / Geoffrey Williamson.
p. cm.
Includes bibliographical references and index.
ISBN 978-1-64728-333-9
1. Industrial engineering. 2. Engineering. 3. Production engineering. I. Williamson, Geoffrey.

T56 .I53 2022
670--dc23

For information on all Willford Press publications
visit our website at www.willfordpress.com

Willford Press

TABLE OF CONTENTS

TABLE OF CONTENTS

PREFACE

This book is a culmination of my many years of practice in this field. I attribute the success of this book to my support group. I would like to thank my parents who have showered me with unconditional love and support and my peers and professors for their constant guidance.

The branch of engineering which is concerned with the optimization of complex systems, organizations and processes is referred to as industrial engineering. This is achieved through improving, developing and implementing the integrated systems of people, information, money, knowledge, energy and materials. Industrial engineering overlaps with various other areas of engineering such as manufacturing engineering, operations engineering, ergonomics, safety engineering, systems engineering, etc. The aim of this field is to specify, predict and evaluate results achieved through industrial processes and systems. Some of the significant areas within this field are production engineering, systems engineering, human factors and safety engineering, etc. This book attempts to understand the multiple branches that fall under the discipline of industrial engineering and how such concepts have practical applications. Most of the topics introduced herein cover new techniques and the applications of this discipline. This book, with its detailed analyses and data, will prove immensely beneficial to professionals and students involved in this area at various levels.

The details of chapters are provided below for a progressive learning:

Chapter - What is Industrial Engineering?

The engineering profession which deals with the optimization of complex systems, processes and organizations is known as industrial engineering. It includes the improvement, development and implementation of the integrated systems of knowledge, people, information, equipment, energy

and material. This is an introductory chapter which will provide a brief introduction to the significant aspects of industrial engineering and manufacturing.

Chapter - Industrial Engineering: Concepts and Practices

Some of the fundamental concepts and practices in industrial engineering include industrial design, production planning, production scheduling, material requirements planning, packaging and labeling. All these concepts and practices of industrial engineering have been carefully analyzed in this chapter.

Chapter - Casting

The engineering manufacturing process in which materials in a molten state are poured into a mold where they solidify is known as casting. This chapter has been carefully written to provide an easy understanding of the process of casting as well as its diverse types such as centrifugal casting, investment casting and sand casting.

Chapter - Molding: Types and Processes

The process of manufacturing which includes the shaping of a liquid or pliable raw material with a rigid frame is known as the molding process. Some of the various types of molding processes include compression molding, blow molding, dip molding, rotational molding, thermoforming, etc. The topics elaborated in this chapter will help in gaining a better perspective about these molding processes.

Chapter - Forming Processes

Forming is a manufacturing process where the shape of a partly finished product is changed using plastic deformation. It includes processes such as tube beading, forging, stamping, shearing, hydroforming, decambering, hemming, explosive forming, etc. This chapter discusses in detail these processes related to forming.

Chapter - Machining

Machining refers to the process in which a piece of raw material is cut into a desired final shape and size by a controlled material-removal process. Some of the machining processes are milling, turning, drilling, broaching, mass finishing, honing and hobbing. These diverse machining processes have been thoroughly discussed in this chapter.

Geoffrey Williamson

What is Industrial Engineering?

CHAPTER 1

The engineering profession which deals with the optimization of complex systems, processes and organizations is known as industrial engineering. It includes the improvement, development and implementation of the integrated systems of knowledge, people, information, equipment, energy and material. This is an introductory chapter which will provide a brief introduction to the significant aspects of industrial engineering and manufacturing.

Industrial engineering is the branch of engineering concerned with the development, improvement, implementation and evaluation of integrated systems of people, money, knowledge, information, equipment, energy, material and process. There are a number of things industrial engineers do in their work to make processes more efficient, to make products more manufacturable and consistent in their quality, and to increase productivity.

Industrial engineering draws upon the principles and methods of engineering analysis and synthesis, as well as mathematical, physical sciences and social sciences together with the principles and methods of engineering analysis and design to specify, predict and evaluate the results to be obtained from such systems. In lean manufacturing systems, industrial engineers work to eliminate wastes of time, money, materials, energy and other resources.

Industrial engineering is also known as operations management, systems engineering, production engineering, manufacturing engineering or manufacturing systems engineering—a distinction that seems to depend on the viewpoint or motives of the user. Recruiters or educational establishments use the names to differentiate themselves from others. In health care, industrial engineers are more commonly known as management engineers, engineering management, or even health systems engineers.

The name "industrial engineer" can be misleading. While the term originally applied to manufacturing, it has grown to encompass services and other industries as well. Similar fields include operations research, systems engineering, ergonomics, process engineering and quality engineering.

Whereas most engineering disciplines apply skills to very specific areas, industrial engineering is applied in virtually every industry. Examples of where industrial engineering might be used include: shortening lines (or queueing theory) at a theme park, streamlining an operating room, distributing products worldwide (also referred to as supply chain management), and manufacturing cheaper and more reliable automobiles. Industrial engineers typically use computer simulation, especially discrete event simulation, for system analysis and evaluation.

Areas of Expertise

The expertise required by an industrial engineer will include some or all of the following elements:

- On demand
  - Investigate problems relating to component quality or difficulties in meeting design and method constraints.
  - Investigate problems with the performance of processes or machines.
  - Implement design changes at the appropriate times.
- Specifically per product (short term)
  - Analysis of the complete product design to determine the way the whole process should be split into steps, or operations, and whether to produce sub-assemblies at certain points in the whole process. This requires knowledge of the facilities available in-house or at sub-contractors.
  - Specification of the method to be used to manufacture or assemble the product(s) at each operation. This includes the machines, tooling, jigs and fixtures and safety equipment, which may have to be designed and built. Notice may need to be taken of any quality procedures and constraints, such as ISO9000. This requires knowledge of health and safety responsibilities and quality policies. This may also involve the creation of programs for any automated machinery.
  - Measurement or calculation of the time required to perform the specified method, taking account of the skills of the operator. This is used to determine the cost the operation performed, to allow balancing of assembly or machining flow lines or the assessment

of the manufacturing capacity required. This technique is known as work study or time and motion studies. These times are also used in value analysis.

- Specification of the storage, handling and transportation methods and equipment required for components and finished product, and at any intermediate stages throughout the whole process. This should eliminate the possibility for damage and minimize the space required.

- Specifically per process (medium term)
  - Determine the maintenance plan for that process.
  - Assess the range of products passing through the process, then investigate the opportunities for process improvement through a reconfiguration of the existing facilities or through the purchase of more efficient equipment. This may also include the out-sourcing of that process. This requires knowledge of design techniques and of investment analysis.
  - Review the individual products passing through the process to identify improvements that can be made by redesign of the product, to reduce (or eliminate) the cost that process adds, or to standardize the components, tooling or methods used.
- Generically (long term)
  - Analyze the flow of products through the facilities of the factory to assess the overall efficiency, and whether the most important products have priority for the most efficient process or machine. This means maximizing throughout for the most profitable products. This requires knowledge of statistical analysis and queuing theory, and of facilities positional layout.
  - Training of new workers in the techniques required to operate the machines or assembly processes.
  - Project planning to achieve timely introduction of new products and processes or changes to them.
  - Generally, a good understanding of the structure and operation of the wider elements of the company, such as sales, purchasing, planning, design and finance; including good communication skills. Modern practice also requires good skills in participation in multi-disciplinary teams.

Value Engineering

Value engineering is based on the proposition that in any complex product, 80 percent of the customers need 20 percent of the features. By focusing on product development, one can produce a superior product at a lower cost for the major part of a market. When a customer needs more features, they are sold to them as options. This approach is valuable in complex electromechanical products such as computer printers, in which the engineering is a major product cost.

To reduce a project's engineering and design costs, it is frequently factored into subassemblies that are designed and developed once and reused in many slightly different products. For example, a typical tape-player has a precision injection-molded tape-deck produced, assembled and tested by a small factory, and sold to numerous larger companies as a subassembly. The tooling and design expense for the tape deck is shared over many products that can look quite different. All that the other products need are the necessary mounting holes and electrical interface.

Quality Control and Quality Assurance

Quality control is a set of measures taken to ensure that defective products or services are not produced, and that the design meets performance requirements. Quality assurance covers all activities from design, development, production, installation, servicing and documentation. This field introduced the rules "fit for purpose" and "do it right the first time."

It is a truism that "quality is free"—very often, it costs no more to produce a product that always works, every time it comes off the assembly line. While this requires a conscious effort during engineering, it can considerably reduce the cost of waste and rework.

Commercial quality efforts have two foci. The first is to reduce the mechanical precision needed to obtain good performance. The second is to control all manufacturing operations to ensure that every part and assembly stays within a specified tolerance.

Statistical process control in manufacturing usually proceeds by randomly sampling and testing a fraction of the output. Testing every output is generally avoided due to time or cost constraints, or because it may destroy the object being tested (such as lighting matches). The variances of critical tolerances are continuously tracked, and manufacturing processes are corrected before bad parts can be produced.

A valuable process industrial engineers perform on a wholly assembled consumer product is called the "shake and bake." Every so often, a whole product is mounted on a shake table in an environmental oven, and operated under increasing vibration, temperatures and humidity until it fails. This finds many unanticipated weaknesses in a product. Another related technique is to operate samples of products until they fail. Generally the data is used to drive engineering and manufacturing process improvements. Often quite simple changes can dramatically improve product service, such as changing to mold-resistant paint, or adding lock-washed placement to the training for new assembly personnel.

Many organizations use statistical process control to bring the organization to "six sigma" levels of quality. In a six sigma organization every item that creates customer value or dissatisfaction is controlled, such that a standard for failure of fewer than four parts in one million is upheld. Items controlled often include clerical tasks such as order-entry, as well as conventional manufacturing processes.

Producibility

Quite frequently, manufactured products have unnecessary precision, production operations, or parts. Simple redesign can eliminate these, lowering costs and increasing manufacturability, reliability and profits.

For example, Russian liquid-fuel rocket motors are intentionally designed to permit ugly (though leak-free) welding, to eliminate grinding and finishing operations that do not help the motor function better.

Another example: rather than unnecessarily requiring parts to be made to extremely precise measurements, some Japanese disc brakes have parts toleranced to three millimeters, an easy-to-meet precision. Yet when combined with crude statistical process controls, this assures that less than one in a million parts will fail to fit.

Many vehicle manufacturers have active programs to reduce the numbers and types of fasteners in their product, to reduce inventory, tooling and assembly costs.

Another producibility technique is "near net shape forming." Often, hundreds of low-precision machining or drilling steps can be eliminated through a premium forming process. For example, precision transfer stamping can quickly produce hundreds of high quality parts from generic rolls of steel and aluminum. Alternatively, die casting can produce metal parts from aluminum or sturdy tin alloys, which are often about as strong as mild steels. Plastic injection molding is another powerful forming technique, especially if the special properties of the part are supplemented with inserts of brass or steel.

When a product incorporates a computer, it replaces many parts with software that fits into a single light-weight, low-power memory part or micro-controller. As computers grow faster, digital signal processing software is beginning to replace many analog electronic circuits for audio and sometimes radio frequency processing.

On some printed circuit boards itself a producibility technique the electrical conduction are intentionally sized to act as delay lines, resistors and inductors to reduce the parts count. An important recent innovation was the use of "surface mounted" components. At one stroke, this eliminated the need to drill most holes in a printed circuit board, as well as clip off the leads after soldering.

In Japan, it is a standard process to design printed circuit boards of inexpensive phenolic resin and paper, and reduce the number of copper layers to one or two to lower costs without harming specifications.

It is becoming increasingly common to consider producibility in the initial stages of product design, a process referred to as design for manufacturability. It is much cheaper to consider these changes during the initial stages of design rather than redesign products after their initial design is complete.

Industrial Engineering Benefits

Industrial engineering has provided a systematic approach to streamline and improve productivity and efficiency. Benefits that can be linked directly to the work of industrial engineers include:

- Leaner, more efficient, and more profitable business practices while increasing customer service and quality.
- Improved efficiency. This improves competitiveness, profitability, and reduces resource requirements.

- The idea of setting labor or time standards. The original production lines in the 1920s were successful because of IEs. The IE profession is timeless and can be molded to fit the times and the place.
- Good organization and improving productivity - these improvements eliminate or reduce some of the frustrations of life and are essential to the long term health of business.
- Increased ability to do more with less.
- Making work safer, faster, easier, and more rewarding.
- Providing a method by which businesses can analyze their processes and try to make improvements to them. It is focused on optimization - doing more with less - and helps to reduce waste in society.
- Increased cycle time and throughput thus helping more people get their product quicker.
- Assistance in guiding society and business to care more for their workforce while improving the bottom-line.
- Showing ways to improve the working environment, improving efficiencies, and teaching people about ergonomics.
- Making the world safer through better designed and easier to use products.
- Reducing costs associated with new technologies, thus allowing more of the population to better their lives by being able to afford technological advances.

MANUFACTURING PROCESS

Manufacturing process is basically a complex activity, concerned with people who've a broad number of disciplines and expertise and a wide range of machinery, tools, and equipment with numerous levels of automation, such as computers, robots, and other equipment. Manufacturing pursuits must be receptive to several needs and developments.

Beside above, all the future technicians must understand the basic needs of workshop routines in terms of man, equipment, material, methods, revenue and other infrastructure conveniences needed to be placed properly for maximum shop or plant layouts and other support solutions effectively regulated or positioned in the field or industry within a properly planned manufacturing firm.

The complete knowledge of fundamental workshop technology and manufacturing processes is highly troublesome for anybody to claim competence over it. It deals with numerous aspects of workshops procedures also for providing the basic working awareness of the various engineering materials, tools, accessories, manufacturing processes, basic concepts of machine instruments, production criteria's, traits and uses of numerous testing instruments and calibrating or inspecting units for checking materials or products designed in various production shops in a commercial environment. It also explains and illustrates the use of several hand tools (calibrating, marking, forming and supporting gear etc.), tools, machinery and diverse methods of production that facilitate forming or shaping the existing raw materials into appropriate usable forms. Below are some of the manufacturing processes that are worth reading.

Types of Manufacturing Processes

Following are the 4 types of manufacturing processes.

Machining

Tools used for machining are immobile power-driven units used to form or shape solid materials, specifically metals. The forming is done by removing extra materials from a work-piece. Machine tools make up the foundation of advanced industry and are utilized either indirectly or directly in the manufacturing of tool parts.

They are categorized under three main categories:

- Traditional Chip-making tools.
- Presses.
- Modern machine tools.

Traditional chip-making tools form the work-piece by trimming away the unwanted part accessible as chips. Presses implement a several shaping processes, which includes shearing, pressing, or elongating. Non-traditional machine tools implement light, electric powered, chemical, and sonic power; superheated gas; and high-energy compound beams to form the exotic supplies and materials that have been created to meet the requirements of modern technology.

Joining

Every joining approach has particular design needs, while certain joint needs may propose a particular joining approach. Design for assembly, and fastener selection apply their own specifications.

Bolting is a standard fastening method, for instance, but welding may cut down the weight of assemblies. Naturally, joints intended for the two approaches would differ tremendously.

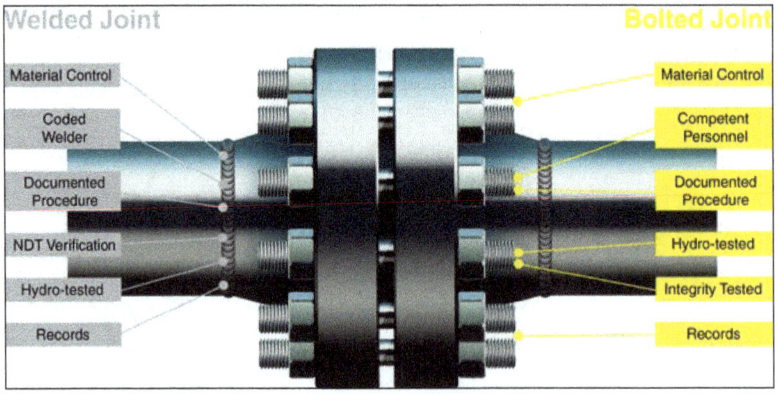

However, all joint patterns must consider features such as load factors, assembly effectiveness, operating surroundings, overhaul and upkeep, and the materials chosen.

Welding is generally a cost-effective approach to fabricate. It doesn't require overlapping materials, and so it removes excess weight brought on by other fastening methods. Fasteners don't have to be purchased and stored in stock. Welding also can minimize costs related to extra parts, for example angles mounted between parts.

Forming

Metal forming is the approach of creating the metallic components by deforming the metal but not by removing, cutting, shredding or breaking any part. Bending, spinning, drawing, and stretching are a few important metal forming process in manufacturing. The metal press such as die and punching tools are implemented for this manufacturing process.

Advantages: Same equipment can be utilized for manufacturing various components by simply changing the dies.

Disadvantages: High apparatus and tooling expenses. Heat treatment must be applied afterwards.

Casting

Casting is a manufacturing process in which a solid is dissolved into a liquid, heated to appropriate temperature (sometimes processed to change its chemical formula), and is then added into a mold or cavity. Thus, in just one step, complex or simple shapes can be crafted from any kind of metal that has the capability to be melted. The end product can have practically any arrangement the designer wants.

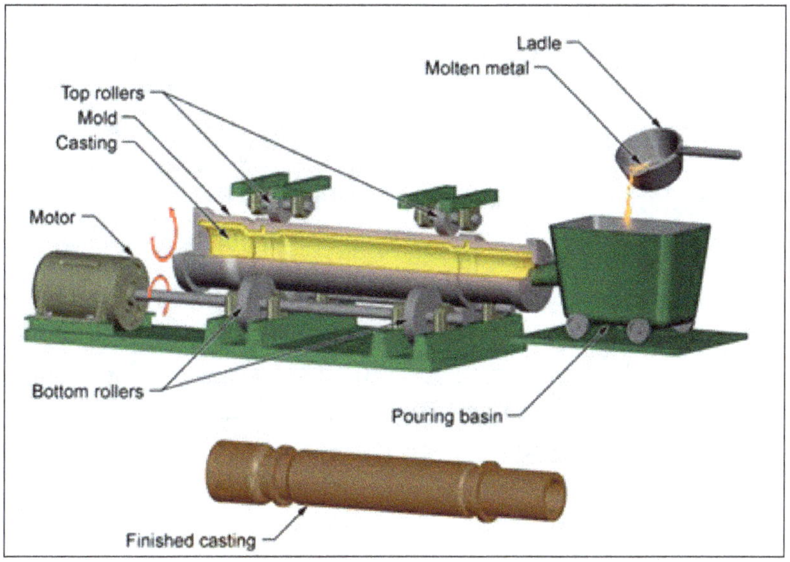

Furthermore, the reluctance to working challenges can be improved, directional attributes can be managed, and a pleasing look can be developed.

Importance of the Manufacturing

Manufacturing is an important process because:

1. Manufactured goods are necessary for trade. According to the World Trade Organization, 80% of interregional trade is in goods, and only 20% is in services. For the U.S., the statistics are about the same. That means that we need goods to trade for foreign goods, or we rack up a large and growing trade deficit, which the United States has been doing for many decades now. This will eventually threaten the value of the dollar; if the dollar becomes very cheap, imports will become very expensive, and the U.S. won't have the capacity to replace imports. In addition, the global trade system has become very unbalanced, with many nations basing their own growth on growth of exports to the U.S., even though they keep taking dollars instead of goods.

2. Manufactured goods are crucial for the service industries. Even though about two-thirds of most economies are composed of service industries, these service industries are dependent on manufactured goods for their operation and for their own technological progress. For instance, the retail and warehousing industries, which comprise about 11 percent of American GNP (value-added), are in the business of selling manufactured goods. The airline industry, the telecommunications industry, and the software industry depend on airplanes, phones and broadcast equipment, and computers for both their existence and for their technological progress.

3. Each manufacturing job creates three other jobs. In the U.S., the Economic Policy Institute has found that each manufacturing job supports three other jobs in the wider economy, through something called "the multiplier effect." That is, the wages from manufacturing employees are re-spent in other parts of the economy, because manufacturing adds so much value to the economy.

4. Economic growth depends on manufacturing. Manufacturing productivity, that is, the goods that are output from a specific amount of input, increases by about 3 percent each year in the U.S., year in and year out, because technological advances are always being made for factory machinery. By contrast, service industries either have very slow productivity growth or depend, directly or indirectly, on technological progress in machinery. In addition, since machines can make other machines, what is called exponential growth, as in quickly reproducing animal populations, can take place.
5. National power depends to a great extent on manufacturing power. Over the last 100 years, the "Great Powers", or most powerful four or five countries, have controlled about 75 percent of global industrial machinery production.This is because industrial machinery is used both to generate national wealth and to produce military equipment. If all regions of the world had an independent capacity to produce manufactured goods, there would be little opportunity to intimidate and dominate countries. In fact, there would probably be fewer wars because global power would be balanced.
6. A world in which all regions had a strong manufacturing base would go far to eliminate poverty and war. Manufacturing creates middle class jobs that anchor a middle class economy. Unions thrive in manufacturing industries because it is easier for the employees to bargain. If all global regions have the power to create the wealth that comes with manufacturing, there will be less opportunity for wars to break out as a result of imbalances of power.

Manufacturing Process Determine Enterprise Profit

Essentially, the purpose of an enterprise is to get profit to the highest number possible. Enterprise will do strategies to maintain the profit. The basic strategy of maintaining profit in manufacturing industry is to make production cost as minimum as possible. This strategy has strong correlation in manufacturing process.

In manufacturing enterprise, there is procedure to conduct manufacturing process. The procedure contains of design the product, selling strategy, and make the product. Product design is one of the factors which need to be well-planned. The design process has huge role in determining the right material with the characteristic of high quality at reasonable price. Designer must be precise to choose the right material that will be added and processed. In example in cellphone manufacture design process, the first step of a designer must consider is the main material the cellphone will use. Is the main material going to be plastic based body or metal based body. When determining which material will be used, designer have to consider which material has better quality with the profitable price.

After product design, enterprise must consider the selling strategy. Selling strategy contains of market status quo, the market demand, and economic environment. If enterprise sell the product against the status quo in the market, the response from the market will not be the desired response due to the buying power, product life cycle, and economic stability which surround the market. For example, an enterprise decided to sell touchscreen cellphone with youth segment as the targeted market. But what happen when the buying power of the market is not as strong as it was expected? assume the targeted country is in economic crisis, the product selling rate will not be as expected as before. The product life cycle also affect the selling, when product approach the decline state of product life cycle, enterprise must innovate to maintain the product life cycle. For example, an enterprise throw a product of qwerty keypad cellphone, in status quo, the market share of cellphone is mainly owned by touchscreen phone, this shows that the market product life cycle is not suitable for selling qwerty keypad cellphone.

After consider the selling strategy, enterprise will make the product and conduct the manufacturing process. The decision need to be done is make or buy. Calculation in manufacturing process is critical in determining the profit enterprise will get. From the case of cellphone design above, assume the designer decided to use metal as the main cellphone body, in manufacturing process, Industrial engineer must consider whether to buy or to make the metal. The calculation of which decision will be taken will affect the production cost. For example, comparison of the price if the engineer buy the metal for IDR 2.500.000 for one cellphone and IDR 2.000.000 if the engineer decided to make the metal. The engineer must consider the other expense such as the supply chain, labor, and raw material availability. If the location does not support the supply chain of raw material and labor availability are limited around the location, then the decision to buy the main cellphone body is the right decision. If the engineer still manage to choose to make it, then what about the other expenses such as additional price which must be paid when delivering the raw material. The standard operational procedure in manufacturing process must be well-excecuted, if not then there will be no standarized product output. When there is no standarized product output, the product quality will be questionable. This condition affect the calculation of the procuction cost by making the calculation not standarized between product output 1 and output 2.

Manufacturing Process Affect the Customer Satifsfaction

Customer satisfaction is a key factor in formation of customer's desires for future purchase. When customer is satisfied by the product, the probability of the customer to buy the product in the future and to promote the product are higher. Many belief that people buy a product based on the recommendation of people around them. It can be parents, friends, or even celebrity who get paid to promote the product that will lead to people to buy the product because they see that the product is recommended. Manufacturing process' role in this factor comes when people examine the product. Does the product comes in smooth process or comes in undesirable process.

When doing manufacturing process, the material must be well-processed in order to prevent from rejected product. Quality of the processed product which will be used by customer must be prioritized. In cellphone manufacturing case, if the assembly process between main body and the screen is not precised, then it will be hard to operate the cellphone and probability of the product cause harm to the customer is higher. When the unqualified product comes to the customer hands, customer considered the product is rejected and customer will be dissapointed because the product is not as expected as before.

Customer satisfaction is an investment, the obstacle of unqualified product output can be maintained by conducting the manufacturing process based on the standard operational procedure to prevent from unqualified product that will lead to customer satisfaction rate decrease.

Manufacturing Process Affect Supply Chain Management

Supply Chain Management (SCM) is a concept that is gaining in popularity and importance. From a practitioner point of view, an Accenture report (in co-operation with Stanford and Insead) states that SCM is critically important or very important to 89% of the surveyed executives. Furthermore, SCM is increasing in importance as 51% of the executives stated that their investments in SCM have increased significantly over the last three years. Since its introduction in the early 1980s, supply chain management (SCM) has become one of the most popular concepts within management in general and within logistics in particular. Supply chain management constitutes the series of interdependent upstream, manufacturing and downstream processes targeted at transforming raw materials into products to meet customer demand.

Industrial Engineering: Concepts and Practices

2 CHAPTER

Some of the fundamental concepts and practices in industrial engineering include industrial design, production planning, production scheduling, material requirements planning, packaging and labeling. All these concepts and practices of industrial engineering have been carefully analyzed in this chapter.

INDUSTRIAL DESIGN

Industrial Design (ID) is the professional practice of designing products and services used by millions of people around the world every day.

Industrial designers typically focus on the appearance, functionality and manufacturability of a product, though they are often involved in far more during a development cycle. All of this ultimately extends to the overall lasting value and experience a product or service provides for end-users.

Every product that you interact with on a daily basis in your home, office, school, or public setting is the result of a design process during which thousands of decisions were made by an industrial designer (and their team) aimed at improving your life through well-executed design.

Industrial Design makes its most significant contribution in differentiation of similar products either by making them appear very different through colour and style, or by enhancing them with detail and minor additional features. In more radical innovation where new technology is involved, industrial design usually has a less central role because the novelty is provided by the new functionality. There are, of course, occasions where the new technology needs to be made more palatable to consumers and then the role of industrial design is more centre stage. An example of this is in areas of medical technology, perhaps where patients need to be re-assured about new machines and procedures.

If we return to the major contribution of industrial design, examples include the increasing variety in lap top computers, the increasing number of small consumer electronics for the home where colour and style are becoming the major differentiators. All increase market range and therefore consumer choice; they don't, however, create a really new market sector. However, a product that did achieve this was the "My First Sony" product. This product was a re-packaging of Sony products, for very young children. The Sony Walkman, was re-created in bright colours with larger and more robust controls. Other products, followed, but this created a new market niche in electronic personal products for very young children. More of a market innovation than a technology one, this is exactly the place where industrial design can be profitably employed.

In these ways the use of industrial design does make a contribution to economic performance and companies, investors and governments need to understand the importance of design spending for the innovation process. Design expenditure ranks with R&D expenditure as a strategic investment for companies, however, the lack of sophisticated techniques applied to design spending often holds back design investment. Project appraisal techniques and issues of finance and accounting methods need consideration. For example, the finance techniques for valuing options used routinely in financial markets are rarely applied in the broader context. The benefits of design spending are systematically undervalued across companies and therefore the economy as a whole.

Decisions on new product development are rightly made at board level in companies. But many executives lack the knowledge of industrial design and this makes it difficult to create a common vision of new product strategy particularly among executives with different backgrounds. Indeed, the need for integration of knowledge is not often emphasised, with the result that a less than optimal strategy is frequently adopted.

The Effects of Organisation Structure on Design

We need to improve the integration of differing design realities, but the structure of most companies, in regard to industrial design makes this quite difficult.

If industrial design comes within a marketing function and marketers buy the creative services of an industrial design consultant on an occasional basis for a special project, we will consider this quite differently than if industrial design is a part of the manufacturing function and we have a couple of engineers working in a small studio turning out tooling design on a day to day basis.

There are three main structural forms for design, one is where a single function has control, another is where there are separate design functions, usually industrial design and engineering, and a third where one of these separate functions dominates the other. A fourth form of structure, though not often seen, is where functions share control of design, for example, marketing and R&D.

The second for, separate design functions is often found in industries with the, "dominant designs" of Abernathy and Utterback, these are mature product types whose attributes time has stabilised so that producers conform to standard configurations – the ball point pen, the automobile, the VCR and increasingly, perhaps, the laptop computer. Here, the structure is decomposed into functions of design. This makes it simpler to manage an ongoing series of modifications – to make one brand or product a bit prettier or a bit faster etc. In the dominant design industries and the car is a good example, basic design remains stable, and so the different designers know more or less what to expect from one another, and can thus get on with their work, largely in isolation, thereby being able to enhance their specialisations and re-inforce their particular design reality.

This model is often quite comforting to both designers and managers but it is not a model that encourages real innovation. Innovation tends to come out of the separate chimneys. Nevertheless, this form of structure has created some very successful enterprises and should not be entirely dismissed.

In the past, when the flaws in this chimney model were first identified, some firms tried an alternative. This was to decide which aspect of design was the most important and to cede power to this group. For a while ford tried placing industrial design in the dominant role with the industrial designers, or stylists as they are often known in the car industry, developing the initial design with other designers in engineering having to conform to it. However, the reverse situation is probably more often seen, where engineers dominate and the industrial designers are bought in to wrap the engineers' product in some kind of pretty skin that ensures that the whole thing ends up being attractive and, "user friendly".

On balance, making one design group dominant over another unbalances the design reality, and is therefore unlikely to be a sustainable model for success. It is a structural form that is best avoided, because as problems become evident in initial design, the only recourse is to throw them back over the wall for redesign which inevitably results in costly delays.

Structure and Integration of Design

Since there is a general move in all industry to flatter, less hierarchical structures and an emphasis on team and cross functional working, the potential to integrate industrial design is quite real. To achieve integration it is essential to consider what we might refer to as a cooperative design structure. In this situation the absolute structure of the company is now, less influential because a commonly held design reality spans all functions. In this design reality, industrial design specialists are respected for their particular capabilities and in turn, these specialists respect the design expertise of general managers. It is these managers who are hands-on designers of product-market fit.

In this design reality, design is not understood as a particular activity undertaken by a particular individual or function. Rather more it is understood as a sort of umbrella. Under the centre of the umbrella are the specialised design activities, the functions of industrial and engineering design and since these are more visible, they can be referred to as "seen design". Beside them, still under the umbrella, are the activities of those in marketing and production who are linking needs of manufacturing processes and the purchasing customers. Holding all this together is the general manager whose task it is to ensure appropriate levels of cooperation among the various specialised functions and the broad fit of product, process and context. All these activities, where managers take decisions on design in the stream of their other decision making can be referred to as "silent design".

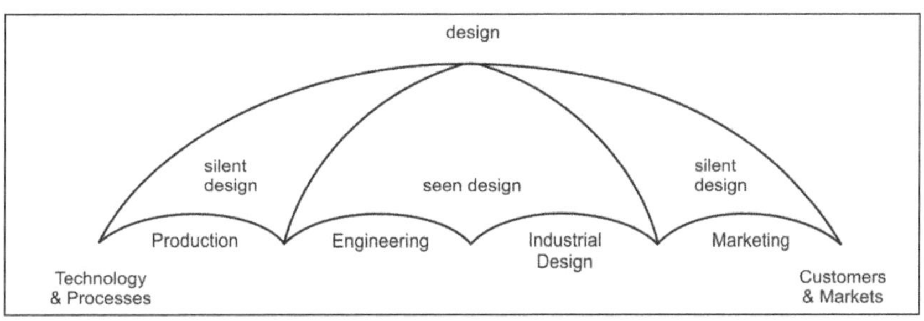

Conceived of in this way, design becomes not simply one activity in the new product development process but the fit of all activities that affect the performance and appearance of the product. It thus emerges as a means to explore and manage various dimensions of design and its various interfaces so that several different realities blend into one integral product.

Adopting the umbrella metaphor provides a fairly unobtrusive way of introducing improvements in industrial design activity as it does not require a protracted period of organisation change. Further, the likelihood of producing an improved product quite quickly is high, which in turn boosts morale and motivation.

Dealing with different realities of design may be uncomfortable initially. But once integration starts, a common design reality can be rich and rewarding and interfaces that were once problematic become opportunities.

PRODUCTION PLANNING

Production planning is the planning of production and manufacturing modules in a company or industry. It utilizes the resource allocation of activities of employees, materials and production capacity, in order to serve different customers.

Different types of production methods, such as single item manufacturing, batch production, mass production, continuous production etc. have their own type of production planning. Production planning can be combined with production control into production planning and control, or it can be combined with enterprise resource planning.

Production planning is used in companies in several different industries, including agriculture industry, oil & gas industry, postal industry, packaging and E-commerce industry, amusement industry, etc.

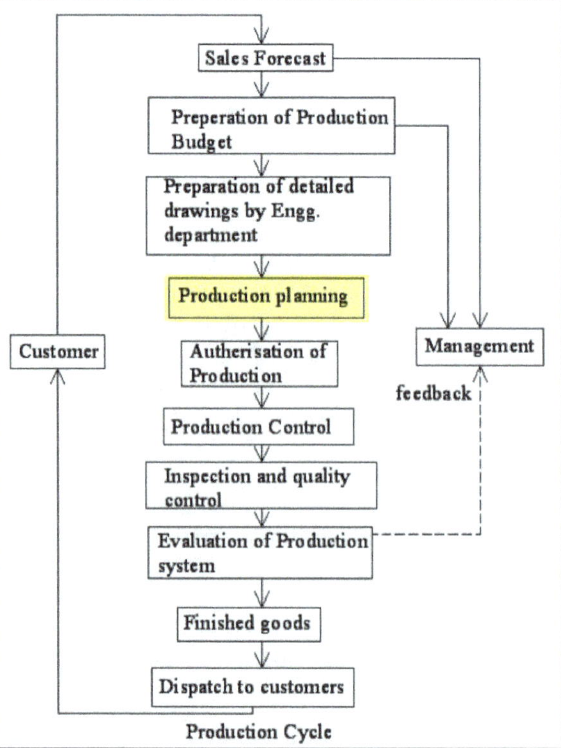

Production Cycle

Role of Production Planning in the Production Cycle.

Production planning is the future of production. It can help in efficient manufacturing or setting up of a production site by facilitating required needs. A production plan is made periodically for a specific time period, called the planning horizon. It can comprise the following activities:

- Determination of the required product mix and factory load to satisfy customers needs.
- Matching the required level of production to the existing resources.
- Scheduling and choosing the actual work to be started in the manufacturing facility.
- Setting up and delivering production orders to production facilities.

In order to develop production plans, the production planner or production planning department needs to work closely together with the marketing department and sales department. They can provide sales forecasts, or a listing of customer orders. The "work is usually selected from a variety of product types which may require different resources and serve different customers. Therefore, the selection must optimize customer-independent performance measures such as cycle time and customer-dependent performance measures such as on-time delivery."

A critical factor in production planning is "the accurate estimation of the productive capacity of available resources, yet this is one of the most difficult tasks to perform well". Production planning should always take "into account material availability, resource availability and knowledge of future demand".

Types of Planning

Different types of production planning can be applied:

- Advanced planning and scheduling.
- Capacity planning.
- Master production schedule.
- Material requirements planning.
- MRP II.
- Scheduling.
- Workflow.

Related kind of planning in organizations:

- Employee scheduling.
- Enterprise resource planning.
- Inventory control.
- Product planning

- Project planning.
- Process planning, redirects to Computer-aided process planning.
- Sales and operations planning.
- Strategy.

Production Control

Production control is the activity of controlling the workflow in the production. It is partly complementary to production planning.

PRODUCTION SCHEDULING

Scheduling is the activity wherein all the resources that are used for the production activities whether it be raw materials, capital, manpower, logistics and any other activity is allocated on a timescale and the timing of production activities is scheduled. It identifies that what resources would be consumed at what stage of production, and according to the estimates a schedule is made so that the company doesn't fall short of resources at the time of production.

The production schedule prepares in-depth estimates of future cash flow also to identify the requirement before taking up a project. Production scheduling tries to optimize the use of the manpower so that there are no excess manpower or shortage of manpower during the production process.

Production scheduling was traditionally done manually using paper and later organizations started using spreadsheets and now a no. of different software's are available for the same. In production scheduling the process usually starts with the identification of the deadline and then moved backward to the current date and in the process the bottleneck processes are identified. Production scheduling takes into account all the constraints like capacity, manpower, Inventory, Plant floor throughput and tries to optimize their use.

Principles of Scheduling

- The principle of optimum task size: Scheduling tends to achieve maximum efficiency when the task sizes are small, and all tasks of same order of magnitude.
- Principle of optimum production plan: The planning should be such that it imposes an equal load on all plants.
- Principle of optimum sequence: Scheduling tends to achieve the maximum efficiency when the work is planned so that work hours are normally used in the same sequence.

Inputs to Scheduling

- Performance standards: The information regarding the performance standards (standard times for operations) helps to know the capacity in order to assign required machine hours to the facility.

- Units in which loading and scheduling is to be expressed.
- Effective capacity of the work centre.
- Demand pattern and extent of flexibility to be provided for rush orders.
- Overlapping of operations.
- Individual job schedules.

Scheduling Strategies

Scheduling strategies vary widely among firms and range from 'no scheduling' to very sophisticated approaches. These strategies are grouped into four classes:

- Detailed scheduling: Detailed scheduling for specific jobs that are arrived from customers is impracticable in actual manufacturing situation. Changes in orders, equipment breakdown, and unforeseen events deviate the plans.
- Cumulative scheduling: Cumulative scheduling of total work load is useful especially for long range planning of capacity needs. This may load the current period excessively and under load future periods. It has some means to control the jobs.
- Cumulative detailed: Cumulative detailed combination is both feasible and practical approach. If master schedule has fixed and flexible portions.
- Priority decision rules: Priority decision rules are scheduling guides that are used independently and in conjunction with one of the above strategies, i.e., first come first serve. These are useful in reducing Work-In-Process (WIP) inventory.

Types of Scheduling

Types of scheduling can be categorized as forward scheduling and backward scheduling.

- Forward scheduling**:** is commonly used in job shops where customers place their orders on "needed as soon as possible" basis. Forward scheduling determines start and finish times of next priority job by assigning it the earliest available time slot and from that time, determines when the job will be finished in that work centre. Since the job and its components start as early as possible, they will typically be completed before they are due at the subsequent work centers in the routing. The forward method generates in the process inventory that are needed at subsequent work centers and higher inventory cost. Forward scheduling is simple to use and it gets jobs done in shorter lead times, compared to backward scheduling.
- Backward scheduling: is often used in assembly type industries and commit in advance to specific delivery dates. Backward scheduling determines the start and finish times for waiting jobs by assigning them to the latest available time slot that will enable each job

to be completed just when it is due, but done before. By assigning jobs as late as possible, backward scheduling minimizes inventories since a job is not completed until it must go directly to the next work centre on its routing. Forward and backward scheduling methods are shown in the following figure.

Forward and Backward Scheduling

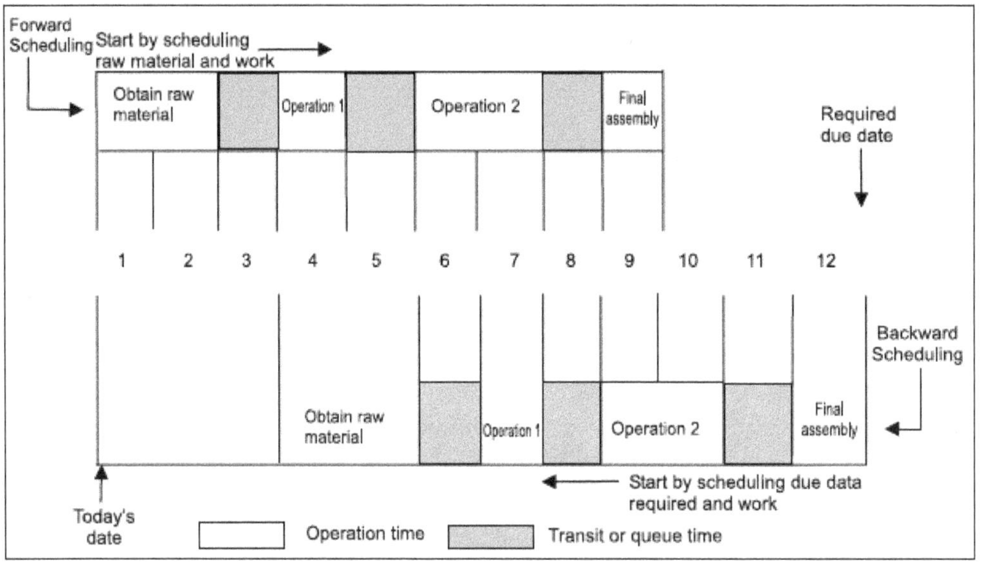

MATERIAL REQUIREMENTS PLANNING

Material requirements planning (MRP) is a production planning, scheduling, and inventory control system used to manage manufacturing processes. Most MRP systems are software-based, but it is possible to conduct MRP by hand as well.

An MRP system is intended to simultaneously meet three objectives:

- Ensure materials are available for production and products are available for delivery to customers.
- Maintain the lowest possible material and product levels in store.
- Plan manufacturing activities, delivery schedules and purchasing activities.

The scope of MRP in Manufacturing

Dependent Demand vs Independent Demand

Independent demand is demand originating outside the plant or production system, while dependent demand is demand for components. The bill of materials (BOM) specifies the relationship between the end product (independent demand) and the components (dependent demand). MRP takes as input the information contained in the BOM.

The basic functions of an MRP system include: inventory control, bill of material processing, and elementary scheduling. MRP helps organizations to maintain low inventory levels. It is used to plan manufacturing, purchasing and delivering activities.

"Manufacturing organizations, whatever their products, face the same daily practical problem - that customers want products to be available in a shorter time than it takes to make them. This means that some level of planning is required."

Companies need to control the types and quantities of materials they purchase, plan which products are to be produced and in what quantities and ensure that they are able to meet current and future customer demand, all at the lowest possible cost. Making a bad decision in any of these areas will make the company lose money. A few examples are given below:

- If a company purchases insufficient quantities of an item used in manufacturing (or the wrong item) it may be unable to meet contract obligations to supply products on time.
- If a company purchases excessive quantities of an item, money is wasted - the excess quantity ties up cash while it remains as stock that might never be used at all.
- Beginning production of an order at the wrong time can cause customer deadlines to be missed.

MRP is a tool to deal with these problems. It provides answers for several questions:

- What items are required?
- How many are required?
- When are they required?

MRP can be applied both to items that are purchased from outside suppliers and to sub-assemblies, produced internally, that are components of more complex items.

Data

The data that must be considered include:

- The end item (or items) being created. This is sometimes called independent demand, or Level "0" on BOM (bill of materials).
- How much is required at a time.
- When the quantities are required to meet demand.
- Shelf life of stored materials.
- Inventory status records. Records of net materials available for use already in stock (on hand) and materials on order from suppliers.
- Bills of materials. Details of the materials, components and sub-assemblies required to make each product.

- Planning data. This includes all the restraints and directions to produce such items as: routing, labor and machine standards, quality and testing standards, pull/work cell and push commands, lot sizing techniques (i.e. fixed lot size, lot-for-lot, economic order quantity), scrap percentages, and other inputs.

Outputs

There are two outputs and a variety of messages/reports:

- Output 1 is the "Recommended Production Schedule." This lays out a detailed schedule of the required minimum start and completion dates, with quantities, for each step of the Routing and Bill Of Material required to satisfy the demand from the master production schedule (MPS).
- Output 2 is the "Recommended Purchasing Schedule." This lays out both the dates on which the purchased items should be received into the facility *and* the dates on which the purchase orders or blanket order release should occur in order to match the production schedules.

Messages and reports:

- Purchase orders. An order to a supplier to provide materials.
- Reschedule notices. These recommend cancelling, increasing, delaying or speeding up existing orders.

Methods to Find Order Quantities

Well-known methods to find order quantities are:

- Dynamic lot-sizing.
- Silver–Meal heuristic.
- Least-unit-cost heuristic.

Mathematical Formulation

MRP can be expressed as an optimal control problem:

Initial conditions:

$$x_i'(0) = x_{i0}' \; i = 1,\ldots.,\text{J} \text{ Dynamics:}$$

$$x_i'(t+1) = x_i'(t) + z_i(t) - d_i'(t) - \sum_{j \in Suc(i)} z_j(t + L_{ij}' - 1)$$

$$t = 0,\ldots,T-1, i = 1,\ldots,J$$

Constraints:

$$x_i'(t) \geq 0 \;\; t = 1,\ldots,T, i = 1,\ldots,J$$

$$z_i(t) \geq 0 \;\; t = 0,\ldots,T-1, i = 1,\ldots,J$$

Objective:

$$min \sum_{i} \sum_{t=0}^{T-1} \left[k_i(t)\delta(z_i(t)) + c_i(t)z_i(t) \right] + \sum_{i} \sum_{t=1}^{T} h_i'(t)x_i'(t)$$

Where *x'* is local inventory (the state), *z* the order size (the control), *d* is local demand, *k* represents fixed order costs, *c* variable order costs, *h* local inventory holding costs. δ() is the Heaviside function. Changing the dynamics of the problem leads to a multi-item analogue of the dynamic lot-size model.

Problems with MRP Systems

- Integrity of the data. If there are any errors in the inventory data, the bill of materials (commonly referred to as 'BOM') data, or the master production schedule, then the output data will also be incorrect ("GIGO": garbage in, garbage out). Data integrity is also affected by inaccurate cycle count adjustments, mistakes in receiving input and shipping output, scrap not reported, waste, damage, box count errors, supplier container count errors, production reporting errors, and system issues. Many of these type of errors can be minimized by implementing pull systems and using bar code scanning. Most vendors in this type of system recommend at least 99% data integrity for the system to give useful results.

- Systems require that the user specify how long it will take for a factory to make a product from its component parts (assuming they are all available). Additionally, the system design also assumes that this "lead time" in manufacturing will be the same each time the item is made, without regard to quantity being made, or other items being made simultaneously in the factory.

- A manufacturer may have factories in different cities or even countries. It is not good for an MRP system to say that we do not need to order some material, because we have plenty of it thousands of miles away. The overall ERP system needs to be able to organize inventory and needs by individual factory and inter-communicate the needs in order to enable each factory to redistribute components to serve the overall enterprise. This means that other systems in the enterprise need to work properly, both before implementing an MRP system and in the future. For example, systems like variety reduction and engineering, which makes sure that product comes out right first time (without defects), must be in place.

- Production may be in progress for some part, whose design gets changed, with customer orders in the system for both the old design, and the new one, concurrently. The overall ERP system needs to have a system of coding parts such that the MRP will correctly calculate needs and tracking for both versions. Parts must be booked into and out of stores more regularly than the MRP calculations take place. Note, these other systems can well be manual systems, but must interface to the MRP. For example, a 'walk around' stock intake done just prior to the MRP calculations can be a practical solution for a small inventory (especially if it is an "open store").

- The other major drawback of MRP is that it fails to account for capacity in its calculations. This means it will give results that are impossible to implement due to manpower, machine or supplier capacity constraints. However this is largely dealt with by MRP II. Generally, MRP II refers to a system with integrated financials. An MRP II system can include finite or infinite capacity planning. But, to be considered a true MRP II system must also include financials. In the MRP II (or MRP2) concept, fluctuations in forecast data are taken into account by including simulation of the master production schedule, thus creating a long-term control. A more general feature of MRP2 is its extension to purchasing, to marketing and to finance (integration of all the functions of the company), ERP has been the next step.

Solutions to Data Integrity Issues

- Bill of material – The best practice is to physically verify the bill of material either at the production site or by disassembling the product.
- Cycle count – The best practice is to determine why a cycle count that increases or decreases inventory has occurred. Find the root cause and correct the problem from occurring again.
- Scrap reporting – This can be the most difficult area to maintain with any integrity. Start with isolating the scrap by providing scrap bins at the production site and then record the scrap from the bins on a daily basis. One benefit of reviewing the scrap on site is that preventive action can be taken by the engineering group.
- Receiving errors – Manual systems of recording what has been received are error prone. The best practice is to implement the system of receiving by ASN from the supplier. The supplier sends an ASN (advanced shipping notification). When the components are received into the facility, the ASN is processed and then company labels are created for each line item. The labels are affixed to each container and then scanned into the MRP system. Extra labels reveal a shortage from the shipment and too few labels reveal an over shipment. Some companies pay for ASN by reducing the time in processing accounts payable.
- Shipping errors – The container labels are printed from the shipper. The labels are affixed to the containers in a staging area or when they are loaded on the transport.
- Production reporting – The best practice is to use bar code scanning to enter production into inventory. A product that is rejected should be moved to an MRB (material review board) location. Containers that require sorting need to be received in reverse.
- Replenishment – The best replenishment practice is replacement using bar code scanning, or via pull system. Depending upon the complexity of the product, planners can actually order materials using scanning with a min-max system.

Demand Driven MRP

In 2011, the third edition of "Orlicky's Materials Requirements Planning" introduced a new type of MRP called "demand driven MRP" (DDMRP).

Demand driven MRP is a multi-echelon formal planning and execution technique with five distinct components:

- Strategic inventory positioning – The first question of effective inventory management is not, "how much inventory should we have?" Nor is it, "when should we make or buy something?" The most fundamental question to ask in today's manufacturing environments is, "given our system and environment, where should we place inventory to have the best protection?" Inventory is like a break wall to protect boats in a marina from the roughness of incoming waves. Out on the open ocean the break walls have to be 50–100 feet tall, but in a small lake the break walls are only a couple feet tall. In a glassy smooth pond no break wall is necessary.

- Buffer profiles and level – Once the strategically replenished positions are determined, the actual levels of those buffers have to be initially set. Based on several factors, different materials and parts behave differently (but many also behave nearly the same). DDMRP calls for the grouping of parts and materials chosen for strategic replenishment and that behave similarly into "buffer profiles." Buffer profiles take into account important factors including lead time (relative to the environment), variability (demand or supply), whether the part is made or bought or distributed and whether there are significant order multiples involved. These buffer profiles are made up of "zones" that produce a unique buffer picture for each part as their respective individual part traits are applied to the group traits.

- Dynamic adjustments – Over the course of time, group and individual traits can and will change as new suppliers and materials are used, new markets are opened and/or old markets deteriorate and manufacturing capacities and methods change. Dynamic buffer levels allow the company to adapt buffers to group and individual part trait changes over time through the use of several types of adjustments. Thus, as more or less variability is encountered or as a company's strategy changes these buffers adapt and change to fit the environment.

- Demand-driven planning – takes advantage of the sheer computational power of today's hardware and software. It also takes advantage of the new demand-driven or pull-based approaches. When these two elements are combined then there is the best of both worlds; relevant approaches and tools for the way the world works today *and* a system of routine that promotes better and quicker decisions and actions at the planning and execution level.

- Highly visible and collaborative execution – Simply launching purchase orders (POs), manufacturing orders (MOs) and transfer orders (TOs) from any planning system does not end the materials and order management challenge. These POs, MOs and TOs have to be effectively managed to synchronize with the changes that often occur within the "execution horizon." The execution horizon is the time from which a PO, MO or TO is opened until the time it is closed in the system of record. DDMRP defines a modern, integrated and greatly needed system of execution for all part categories in order to speed the proliferation of relevant information and priorities throughout an organization and supply chain.

These five components work together to greatly dampen, if not eliminate, the nervousness of traditional MRP systems and the bullwhip effect in complex and challenging environments. The

Demand Driven Institute claims the following: In utilizing these approaches, planners will no longer have to try to respond to every single message for every single part that is off by even one day. This approach provides real information about those parts that are truly at risk of negatively impacting the planned availability of inventory. DDMRP sorts the significant few items that require attention from the many parts that are being managed. Under the DDMRP approach, fewer planners can make better decisions more quickly. That means companies will be better able to leverage their working and human capital as well as the huge investments they have made in information technology. One down-side, however, is that DDMRP can not run on the majority of MRPII/ERP systems in use today, so organisations will have to implement a compatible system in order to run DDMRP.

DDMRP has been successfully applied to a variety of environments including CTO (configure to order), MTS (make to stock), MTO (make to order) and ETO (engineer to order). The methodology is applied differently in each environments but the five step process remains the same. DDMRP leverages knowledge from theory of constraints (TOC), traditional MRP & DRP, Six Sigma and lean. It is effectively an amalgam of MRP and kanban techniques. As such, it incorporates the strengths of both but also the weaknesses of both; hence its limited adoption.

PACKAGING AND LABELING

Packaging is the science, art and technology of enclosing or protecting products for distribution, storage, sale, and use. Packaging also refers to the process of designing, evaluating, and producing packages. Packaging can be described as a coordinated system of preparing goods for transport, warehousing, logistics, sale, and end use. Packaging contains, protects, preserves, transports, informs, and sells. In many countries it is fully integrated into government, business, institutional, industrial, and personal use.

Package labeling (American English) or labelling (British English) is any written, electronic, or graphic communication on the package or on a separate but associated label.

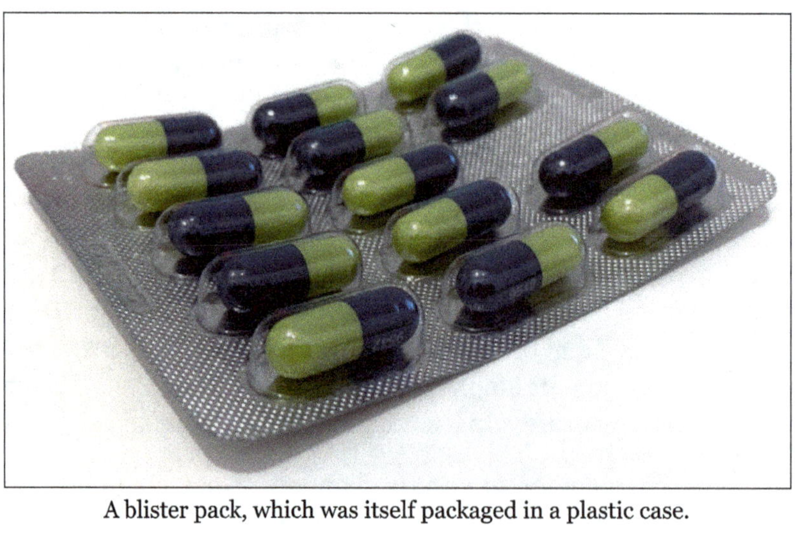

A blister pack, which was itself packaged in a plastic case.

The Purposes of Packaging and Package Labels

Packaging and package labeling have several objectives

- Physical protection – The objects enclosed in the package may require protection from, among other things, mechanical shock, vibration, electrostatic discharge, compression, temperature, etc.

- Barrier protection – A barrier to oxygen, water vapor, dust, etc., is often required. Permeation is a critical factor in design. Some packages contain desiccants or oxygen absorbers to help extend shelf life. Modified atmospheres or controlled atmospheres are also maintained in some food packages. Keeping the contents clean, fresh, sterile and safe for the duration of the intended shelf life is a primary function. A barrier is also implemented in cases where segregation of two materials prior to end use is required, as in the case of special paints, glues, medical fluids, etc.

- Containment or agglomeration – Small objects are typically grouped together in one package for reasons of storage and selling efficiency. For example, a single box of 1000 pencils requires less physical handling than 1000 single pencils. Liquids, powders, and granular materials need containment.

- Information transmission – Packages and labels communicate how to use, transport, recycle, or dispose of the package or product. With pharmaceuticals, food, medical, and chemical products, some types of information are required by government legislation. Some packages and labels also are used for track and trace purposes. Most items include their serial and lot numbers on the packaging, and in the case of food products, medicine, and some chemicals the packaging often contains an expiry/best-before date, usually in a shorthand form. Packages may indicate their construction material with a symbol.

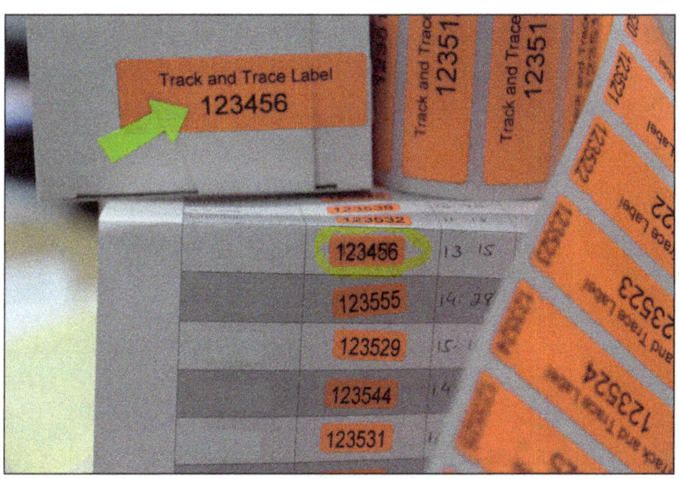

Permanent, tamper evident voiding label with a dual number tab to help keep packaging secure with the additional benefit of being able to track and trace parcels and packages.

- Marketing – Packaging and labels can be used by marketers to encourage potential buyers to purchase a product. Package graphic design and physical design have been important and constantly evolving phenomena for several decades. Marketing communications and graphic design are applied to the surface of the package and often to the point of sale display. Most packaging is designed to reflect the brand's message and identity on the one hand while highlighting the respective product concept on the other hand.

A single-serving shampoo packet.

- Security – Packaging can play an important role in reducing the security risks of shipment. Packages can be made with improved tamper resistance to deter manipulation and they can also have tamper-evident features indicating that tampering has taken place. Packages can be engineered to help reduce the risks of package pilferage or the theft and resale of products. Some package constructions are more resistant to pilferage than other types, and some have pilfer-indicating seals. Counterfeit consumer goods, unauthorized sales (diversion), material substitution and tampering can all be minimized or prevented with such anti-counterfeiting technologies. Packages may include authentication seals and use security printing to help indicate that the package and contents are not counterfeit. Packages also can include anti-theft devices such as dye-packs, RFID tags, or electronic article surveillance tags that can be activated or detected by devices at exit points and require specialized tools to deactivate. Using packaging in this way is a means of retail loss prevention.

- Convenience – Packages can have features that add convenience in distribution, handling, stacking, display, sale, opening, reclosing, using, dispensing, reusing, recycling, and ease of disposal

- Portion control – Single serving or single dosage packaging has a precise amount of contents to control usage. Bulk commodities (such as salt) can be divided into packages that are a more suitable size for individual households. It also aids the control of inventory: selling sealed one-liter bottles of milk, rather than having people bring their own bottles to fill themselves.

- Branding/Positioning – Packaging and labels are increasingly used to go beyond marketing to brand positioning, with the materials used and design chosen key to the storytelling element of brand development. Due to the increasingly fragmented media landscape in the digital age this aspect of packaging is of growing importance.

Packaging Types

Various types of household packaging for foods.

Packaging may be of several different types. For example, a transport package or distribution package can be the shipping container used to ship, store, and handle the product or inner packages. Some identify a consumer package as one which is directed toward a consumer or household.

Packaging may be described in relation to the type of product being packaged: medical device packaging, bulk chemical packaging, over-the-counter drug packaging, retail food packaging, military materiel packaging, pharmaceutical packaging, etc.

It is sometimes convenient to categorize packages by layer or function: "primary", "secondary", etc.

- Primary packaging is the material that first envelops the product and holds it. This usually is the smallest unit of distribution or use and is the package which is in direct contact with the contents.
- Secondary packaging is outside the primary packaging, and may be used to prevent pilferage or to group primary packages together.
- Tertiary or transit packaging is used for bulk handling, warehouse storage and transport shipping. The most common form is a palletized unit load that packs tightly into containers.

These broad categories can be somewhat arbitrary. For example, depending on the use, a shrink wrap can be primary packaging when applied directly to the product, secondary packaging when used to combine smaller packages, or tertiary packaging when used to facilitate some types of distribution, such as to affix a number of cartons on a pallet.

Packaging can also have categories based on the package form. For example, thermoform packaging and flexible packaging describe broad usage areas.

Symbols used on Packages and Labels

Many types of symbols for package labeling are nationally and internationally standardized. For consumer packaging, symbols exist for product certifications (such as the FCC and TÜV marks), trademarks, proof of purchase, etc. Some requirements and symbols exist to communicate aspects of consumer rights and safety, for example the CE marking or the estimated sign that notes conformance to EU weights and measures accuracy regulations. Examples of environmental and recycling symbols include the recycling symbol, the recycling code (which could be a resin identification code), and the "Green Dot". Food packaging may show food contact material symbols. In the European Union, products of animal origin which are intended to be consumed by humans have to carry standard, oval-shaped EC identification and health marks for food safety and quality insurance reasons.

A bar code on a tin of condensed milk.

Bar codes, Universal Product Codes, and RFID labels are common to allow automated information management in logistics and retailing. Country-of-origin labeling is often used. Some products might use QR codes or similar matrix barcodes. Packaging may have visible registration marks and other printing calibration and troubleshooting cues.

Shipping Container Labeling

"Print & Apply" corner wrap UCC (GS1-128) label application to a pallet load.

Technologies related to shipping containers are identification codes, bar codes, and electronic data interchange (EDI). These three core technologies serve to enable the business functions in the process of shipping containers throughout the distribution channel. Each has an essential function: identification codes either relate product information or serve as keys to other data, bar codes allow for the automated input of identification codes and other data, and EDI moves data between trading partners within the distribution channel.

Elements of these core technologies include UPC and EAN item identification codes, the SCC-14 (UPC shipping container code), the SSCC-18 (Serial Shipping Container Codes), Interleaved 2-of-5 and UCC/EAN-128 (newly designated GS1-128) bar code symbologies, and ANSI ASC X12 and UN/EDIFACT EDI standards.

Small parcel carriers often have their own formats. For example, United Parcel Service has a MaxiCode 2-D code for parcel tracking.

RFID labels for shipping containers are also increasingly used. A Wal-Mart division, Sam's Club, has also moved in this direction and is putting pressure on its suppliers to comply.

Shipments of hazardous materials or dangerous goods have special information and symbols (labels, placards, etc.) as required by UN, country, and specific carrier requirements. On transport packages, standardized symbols are also used to communicate handling needs. Some are defined in the ASTM D5445 "Standard Practice for Pictorial Markings for Handling of Goods" and ISO 780 "Pictorial marking for handling of goods".

Flammable liquid.

Explosives.

This way up.

Fragile material.

Keep away from water.

Package Development Considerations

Package design and development are often thought of as an integral part of the new product development process. Alternatively, development of a package (or component) can be a separate process, but must be linked closely with the product to be packaged. Package design starts with the identification of all the requirements: structural design, marketing, shelf life, quality assurance, logistics, legal, regulatory, graphic design, end-use, environmental, etc. The design criteria, performance (specified by package testing), completion time targets, resources, and cost constraints need to be established and agreed upon. Package design processes often employ rapid prototyping, computer aided design, computer-aided manufacturing and document automation.

Transport packaging needs to be matched to its logistics system. Packages designed for controlled shipments of uniform pallet loads may not be suited to mixed shipments with express carriers.

An example of how package design is affected by other factors is its relationship to logistics. When the distribution system includes individual shipments by a small parcel carrier, the sorting, handling, and mixed stacking make severe demands on the strength and protective ability of the transport package. If the logistics system consists of uniform palletized unit loads, the structural design of the package can be designed to meet those specific needs, such as vertical stacking for a longer time frame. A package designed for one mode of shipment may not be suited to another.

With some types of products, the design process involves detailed regulatory requirements for the packaging. For example, any package components that may contact foods are designated food contact materials. Toxicologists and food scientists need to verify that such packaging materials are allowed by applicable regulations. Packaging engineers need to verify that the completed package will keep the product safe for its intended shelf life with normal usage. Packaging processes, labeling, distribution, and sale need to be validated to assure that they comply with regulations that have the well being of the consumer in mind.

Sometimes the objectives of package development seem contradictory. For example, regulations for an over-the-counter drug might require the package to be tamper-evident and child resistant: These intentionally make the package difficult to open. The intended consumer, however, might be handicapped or elderly and unable to readily open the package. Meeting all goals is a challenge.

Package design may take place within a company or with various degrees of external packaging engineering: independent contractors, consultants, vendor evaluations, independent laboratories, contract packagers, total outsourcing, etc. Some sort of formal project planning and project management methodology is required for all but the simplest package design and development programs. An effective quality management system and Verification and Validation protocols are mandatory for some types of packaging and recommended for all.

Environmental Considerations

Package development involves considerations of sustainability, environmental responsibility, and applicable environmental and recycling regulations. It may involve a life cycle assessment which considers the material and energy inputs and outputs to the package, the packaged product, the packaging process, the logistics system, waste management, etc. It is necessary to know the relevant regulatory requirements for point of manufacture, sale, and use.

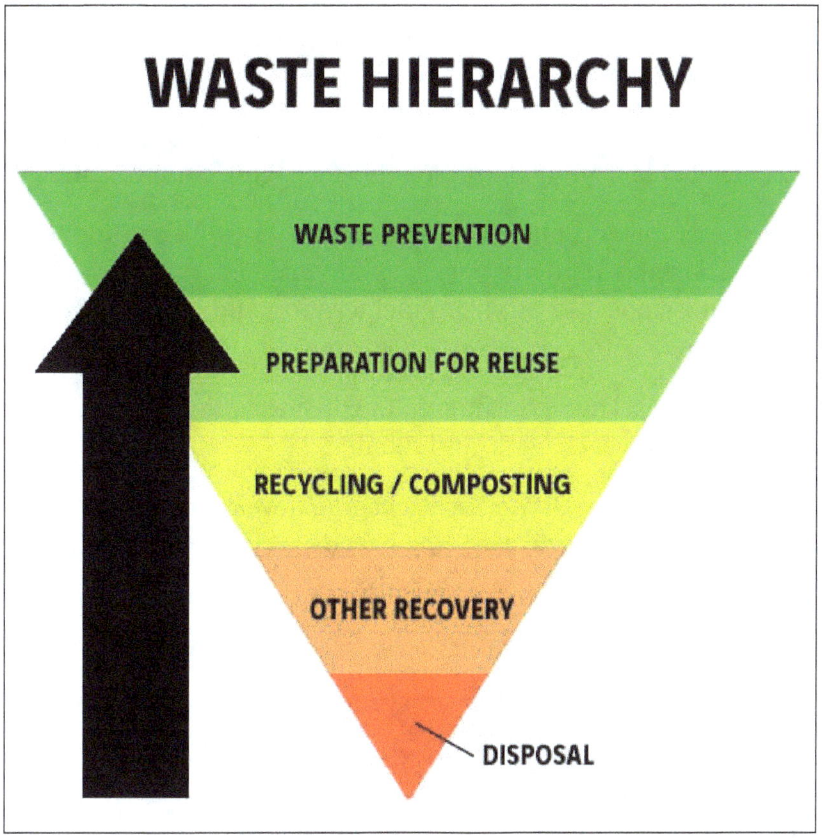

The waste hierarchy.

The traditional "three R's" of reduce, reuse, and recycle are part of a waste hierarchy which may be considered in product and package development.

- Prevention – Waste prevention is a primary goal. Packaging should be used only where needed. Proper packaging can also help prevent waste. Packaging plays an important part in preventing loss or damage to the packaged product. Usually, the energy content and material usage of the product being packaged are much greater than that of the package. A vital function of the package is to protect the product for its intended use: if the product is damaged or degraded, its entire energy and material content may be lost.

- Minimization (also "source reduction") – The mass and volume of packaging (per unit of contents) can be measured and used as criteria for minimizing the package in the design process. Usually "reduced" packaging also helps minimize costs. Packaging engineers continue to work toward reduced packaging.

- Reuse – Reusable packaging is encouraged. Returnable packaging has long been useful (and economically viable) for closed loop logistics systems. Inspection, cleaning, repair and recouperage are often needed. Some manufacturers re-use the packaging of the incoming parts for a product, either as packaging for the outgoing product or as part of the product itself.

- Recycling – Recycling is the reprocessing of materials (pre- and post-consumer) into new products. Emphasis is focused on recycling the largest primary components of a package: steel, aluminum, papers, plastics, etc. Small components can be chosen which are not difficult to separate and do not contaminate recycling operations. Packages can sometimes be designed to separate components to better facilitate recycling.

- Energy recovery – Waste-to-energy and Refuse-derived fuel in approved facilities make use of the heat available from incinerating the packaging components.

- Disposal – Incineration, and placement in a sanitary landfill are undertaken for some materials. Certain US states regulate packages for toxic contents, which have the potential to contaminate emissions and ash from incineration and leachate from landfill. Packages should not be littered.

Development of sustainable packaging is an area of considerable interest to standards organizations, governments, consumers, packagers, and retailers.

Sustainability is the fastest-growing driver for packaging development, particularly for packaging manufacturers that work with the world's leading brands, as their CSR (Corporate Social Responsibility) targets often exceed those of the EU Directive.

Packaging Machinery

Beer bottling lines.

Choosing packaging machinery includes an assessment of technical capabilities, labor requirements, worker safety, maintainability, serviceability, reliability, ability to integrate into the packaging line, capital cost, floorspace, flexibility (change-over, materials, multiple products, etc.), energy requirements, quality of outgoing packages, qualifications (for food, pharmaceuticals, etc.), throughput, efficiency, productivity, ergonomics, return on investment, etc.

Packaging machinery can be:

- Purchased as standard, off-the-shelf equipment.

- Purchased custom-made or custom-tailored to specific operations.

- Manufactured or modified by in-house engineers and maintenance staff.

Efforts at packaging line automation increasingly use programmable logic controllers and robotics.

Packaging machines may be of the following general types:

- Accumulating and collating machines.
- Blister packs, skin packs and vacuum packaging machines.
- Bottle caps equipment, over-capping, lidding, closing, seaming and sealing machines.
- Box, case, tray, and carrier forming, packing, unpacking, closing, and sealing machines.
- Cartoning machines.
- Cleaning, sterilizing, cooling and drying machines.
- Coding, printing, marking, stamping, and imprinting machines.
- Converting machines.
- Conveyor belts, accumulating and related machines.
- Feeding, orienting, placing and related machines.
- Filling machines: handling dry, powdered, solid, liquid, gas, or viscous products.
- Inspecting: visual, sound, metal detecting, etc.
- Label dispenser.
- Orienting, unscrambling machines.
- Package filling and closing machines.
- Palletizing, depalletizing, unit load assembly.
- Product identification: labeling, marking, etc.
- Sealing machines: heat sealer or glue units.
- Slitting machines.
- Weighing machines: check weigher, multihead weigher.
- Wrapping machines: stretch wrapping, shrink wrap, banding.
- Form, fill and seal machines.
- Other specialty machinery: slitters, perforating, laser cutters, parts attachment, etc.

Bakery goods shrinkwrapped by shrink film, heat sealer and heat tunnel on roller conveyor.

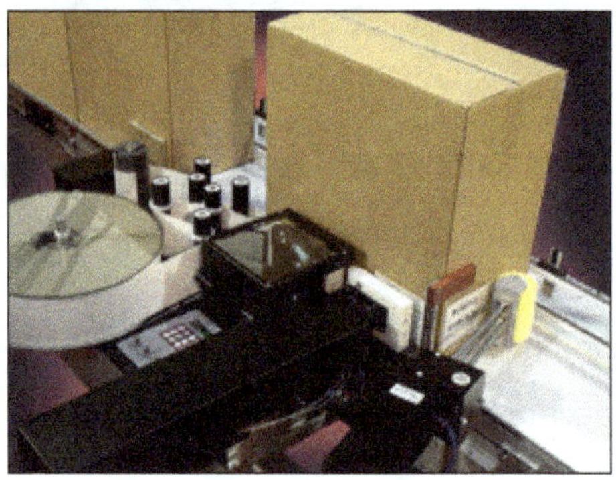

Label printer applicator applying a label to adjacent panels of a corrugated box.

SAFETY ENGINEERING

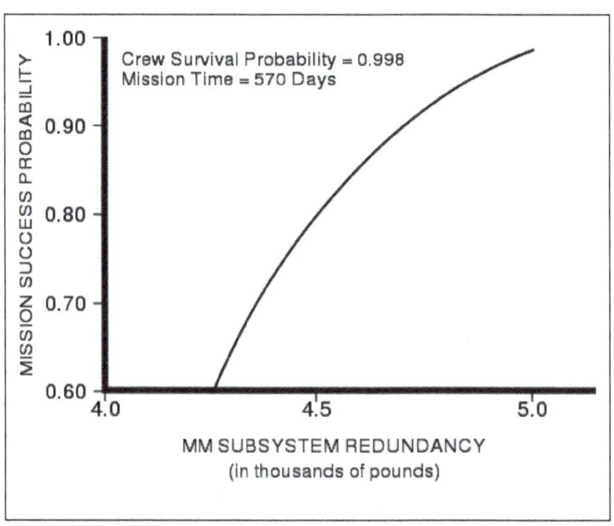

A NASA graph shows the relationship between the survival of a crew of astronauts and the amount of redundant equipment in their spacecraft (the "MM," Mission Module).

Safety engineering is an applied science closely related to systems engineering and its subset,

System Safety Engineering. Safety engineering assures that a life-critical system behaves as needed even when other components fail. In practical terms, the term "safety engineering" refers to any act of accident prevention by a person qualified in the field. Safety engineering is often reactionary to adverse events, also described as "incidents," as reflected in accident statistics. This arises largely because of the complexity and difficulty of collecting and analyzing data on "near misses."

Increasingly, the importance of a safety review is being recognized as an important risk management tool. Failure to identify risks to safety, and the according inability to address or "control" these risks, can result in massive costs, both human and economic. The multidisciplinary nature of safety engineering means that a very broad array of professionals are actively involved in accident prevention or safety engineering.

The Task of Safety Engineers

The majority of those practicing safety engineering are employed in industry to keep workers safe on a day to day basis.

Safety engineers distinguish different extents of defective operation. A failure is "the inability of a system or component to perform its required functions within specified performance requirements," while a fault is "a defect in a device or component, for example: A short circuit or a broken wire". System-level failures are caused by lower-level faults, which are ultimately caused by basic component faults. (Some texts reverse or confuse these two terms.) The unexpected failure of a device that was operating within its design limits is a primary failure, while the expected failure of a component stressed beyond its design limits is a secondary failure. A device which appears to malfunction because it has responded as designed to a bad input is suffering from a command fault.

A critical fault endangers one or a few people. A catastrophic fault endangers, harms, or kills a significant number of people.

Safety engineers also identify different modes of safe operation: A probabilistically safe system has no single point of failure, and enough redundant sensors, computers, and effectors so that it is very unlikely to cause harm (usually "very unlikely" means, on average, less than one human life lost in a billion hours of operation). An inherently safe system is a clever mechanical arrangement that cannot be made to cause harm—obviously the best arrangement, but this is not always possible. A fail-safe system is one that cannot cause harm when it fails. A fault-tolerant system can continue to operate with faults, though its operation may be degraded in some fashion.

These terms combine to describe the safety needed by systems: For example, most biomedical equipment is only "critical," and often another identical piece of equipment is nearby, so it can be merely "probabilistically fail-safe." Train signals can cause "catastrophic" accidents (imagine chemical releases from tank-cars) and are usually "inherently safe." Aircraft "failures" are "catastrophic" (at least for their passengers and crew) so aircraft are usually "probabilistically fault-tolerant." Without any safety features, nuclear reactors might have "catastrophic failures," so real nuclear reactors are required to be at least "probabilistically fail-safe," and some, such as pebble bed reactors, are "inherently fault-tolerant."

The Process

Ideally, safety engineers take an early design of a system, analyze it to find what faults can occur, and then propose safety requirements in design specifications up front and changes to existing systems to make the system safer. In an early design stage, often a fail-safe system can be made acceptably safe with a few sensors and some software to read them. Probabilistic fault-tolerant systems can often be made by using more, but smaller and less-expensive pieces of equipment.

Far too often, rather than actually influencing the design, safety engineers are assigned to prove that an existing, completed design is safe. If a safety engineer then discovers significant safety problems late in the design process, correcting them can be very expensive. This type of error has the potential to waste large sums of money.

The exception to this conventional approach is the way some large government agencies approach safety engineering from a more proactive and proven process perspective. This is known as System Safety. The System Safety philosophy, supported by the System Safety Society and many other organizations, is to be applied to complex and critical systems, such as commercial airliners, military aircraft, munitions and complex weapon systems, spacecraft and space systems, rail and transportation systems, air traffic control system and more complex and safety-critical industrial systems. The proven System Safety methods and techniques are to prevent, eliminate and control hazards and risks through designed influences by a collaboration of key engineering disciplines and product teams. Software safety is fast growing field since modern systems functionality are increasingly being put under control of software. The whole concept of system safety and software safety, as a subset of systems engineering, is to influence safety-critical systems designs by conducting several types of hazard analyses to identify risks and to specify design safety features and procedures to strategically mitigate risk to acceptable levels before the system is certified.

Additionally, failure mitigation can go beyond design recommendations, particularly in the area of maintenance. There is an entire realm of safety and reliability engineering known as "Reliability Centered Maintenance" (RCM), which is a discipline that is a direct result of analyzing potential failures within a system and determining maintenance actions that can mitigate the risk of failure. This methodology is used extensively on aircraft and involves understanding the failure modes of the serviceable replaceable assemblies in addition to the means to detect or predict an impending failure. Every automobile owner is familiar with this concept when they take in their car to have the oil changed or brakes checked. Even filling up one's car with gas is a simple example of a failure mode (failure due to fuel starvation), a means of detection (fuel gauge), and a maintenance action (filling the tank).

For large scale complex systems, hundreds if not thousands of maintenance actions can result from the failure analysis. These maintenance actions are based on conditions (for example, gauge reading or leaky valve), hard conditions (for example, a component is known to fail after 100 hrs of operation with 95 percent certainty), or require inspection to determine the maintenance action (such as metal fatigue). The Reliability Centered Maintenance concept then analyzes each individual maintenance item for its risk contribution to safety, mission, operational readiness, or cost to repair if a failure does occur. Then the sum total of all the maintenance actions are bundled into maintenance intervals so that maintenance is not occurring around the clock, but rather, at regular intervals. This bundling process introduces further complexity, as it might stretch some maintenance cycles, thereby increasing risk, but reduce others, thereby potentially reducing risk, with the end result being a comprehensive maintenance schedule, purpose built to reduce operational risk and ensure acceptable levels of operational readiness and availability.

Analysis Techniques

The two most common fault modeling techniques are called "failure modes and effects analysis" and "fault tree analysis." These techniques are just ways of finding problems and of making plans to cope with failures, as in Probabilistic Risk Assessment (PRA or PSA). One of the earliest complete studies using PRA techniques on a commercial nuclear plant was the Reactor Safety Study (RSS), edited by Prof. Norman Rasmussen.

Failure Modes and Effects Analysis

In the technique known as "failure mode and effects analysis" (FMEA), an engineer starts with a block diagram of a system. The safety engineer then considers what happens if each block of the diagram fails. The engineer then draws up a table in which failures are paired with their effects and an evaluation of the effects. The design of the system is then corrected, and the table adjusted until the system is not known to have unacceptable problems. It is very helpful to have several engineers review the failure modes and effects analysis.

Fault Tree Analysis

First a little history to put FTA into perspective. It came out of work on the Minuteman Missile System. All the digital circuits used in the Minuteman Missile System were designed and tested extensively. The failure probabilities as well as failure modes well understood and documented for each circuit. GTE/Sylvania, one of the prime contractors, discovered that the probability of failure for various components were easily constructed from the Boolean expressions for those components. In any case this analysis of hardware lead to the use of the same symbology and thinking for what (with additional symbols) is now known as a Fault Tree. Note the de Morgan's equivalent of a fault tree is the success tree.

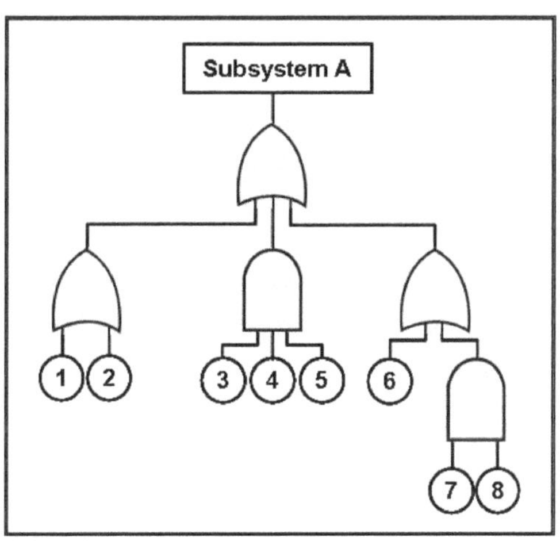

A fault tree diagram.

In the technique known as "fault tree analysis," an undesired effect is taken as the root ('top event') of a tree of logic. There should be *only* one Top Event and all concerns must tree down from it. This is also a consequence of another Minuteman Missile System requirement that all analysis be Top Down. By fiat there was to be no bottom up analysis. Then, each situation that could cause that effect is added to the tree as a series of logic expressions. When fault trees are labeled with actual numbers about failure probabilities, which are often in practice unavailable because of the expense of testing, computer programs can calculate failure probabilities from fault trees.

The Tree is usually written out using conventional logic gate symbols. The route through a Tree between an event and an initiator in the tree is called a Cutset. The shortest credible way through the tree from Fault to initiating Event is called a Minimal Cutset.

Some industries use both Fault Trees and Event Trees. An Event Tree starts from an undesired initiator (loss of critical supply, component failure etc) and follows possible further system events through to a series of final consequences. As each new event is considered, a new node on the tree is added with a split of probabilities of taking either branch. The probabilities of a range of "top events" arising from the initial event can then be seen.

Classic programs include the Electric Power Research Institute's (EPRI) CAFTA software, which is used by almost all the U.S. nuclear power plants and by a majority of U.S. and international aerospace manufacturers, and the Idaho National Laboratory's SAPHIRE, which is used by the U.S. Government to evaluate the safety and reliability of nuclear reactors, the Space Shuttle, and the International Space Station.

Safety Certification

Usually a failure in safety-certified systems is acceptable if, on average, less than one life per 10^9 hours of continuous operation is lost to failure. Most Western nuclear reactors, medical equipment, and commercial aircraft are certified to this level. The cost versus loss of lives has been considered appropriate at this level (by FAA for aircraft under Federal Aviation Regulations).

Preventing Failure

Probabilistic Fault Tolerance: Adding Redundancy to Equipment and Systems

Once a failure mode is identified, it can usually be prevented entirely by adding extra equipment to the system. For example, nuclear reactors contain dangerous radiation, and nuclear reactions can cause so much heat that no substance might contain them. Therefore reactors have emergency core cooling systems to keep the temperature down, shielding to contain the radiation, and engineered barriers (usually several, nested, surmounted by a containment building) to prevent accidental leakage.

Most biological organisms have a certain amount of redundancy: Multiple organs, multiple limbs, and so on.

For any given failure, a fail-over, or redundancy can almost always be designed and incorporated into a system.

When does Safety Stop, where does Reliability begin?

Assume there is a new design for a submarine. In the first case, as the prototype of the submarine is being moved to the testing tank, the main hatch falls off. This would be easily defined as an unreliable hatch. Now the submarine is submerged to 10,000 feet, whereupon the hatch falls off again, and all on board are killed. The failure is the same in both cases, but in the second case it becomes a safety issue. Most people tend to judge risk on the basis of the likelihood of occurrence. Other people judge risk on the basis of their magnitude of regret, and are likely unwilling to accept risk no matter how unlikely the event. The former make good reliability engineers, the latter make good safety engineers.

Perhaps there is a need to design a Humvee with a rocket launcher attached. The reliability engineer could make a good case for installing launch switches all over the vehicle, making it very likely someone can reach one and launch the rocket. The safety engineer could make an equally compelling case for putting only two switches at opposite ends of the vehicle which must both be thrown to launch the rocket, thus ensuring the likelihood of an inadvertent launch was small. An additional irony is that it is unlikely that the two engineers can reconcile their differences, in which case a manager who does not understand the technology could choose one design over the other based on other criteria, like cost of manufacturing.

Inherent Fail-safe Design

When adding equipment is impractical (usually because of expense), then the least expensive form of design is often "inherently fail-safe." The typical approach is to arrange the system so that ordinary single failures cause the mechanism to shut down in a safe way. (For nuclear power plants, this is termed a passively safe design, although more than ordinary failures are covered).

One of the most common fail-safe systems is the overflow tube in baths and kitchen sinks. If the valve sticks open, rather than causing an overflow and damage, the tank spills into an overflow.

Another common example is that in an elevator the cable supporting the car keeps spring-loaded brakes open. If the cable breaks, the brakes grab rails, and the elevator cabin does not fall.

Inherent fail-safes are common in medical equipment, traffic and railway signals, communications equipment, and safety equipment.

Containing Failure

It is also common practice to plan for the failure of safety systems through containment and isolation methods. The use of isolating valves, also known as the Block and bleed manifold, is very common in isolating pumps, tanks, and control valves that may fail or need routine maintenance. In addition, nearly all tanks containing oil or other hazardous chemicals are required to have containment barriers set up around them to contain 100 percent of the volume of the tank in the event of a catastrophic tank failure. Similarly, long pipelines have remote-closing valves periodically installed in the line so that in the event of failure, the entire pipeline is not lost. The goal of all such containment systems is to provide means of limiting the damage done by a failure to a small localized area.

References

- Ptak, Carol; Smith, Chad (2011). Orlicky's Materials Requirements Planning. Mcgraw Hill. ISBN 978-0-07-175563-4
- What-industrial-design: idsa.org, Retrieved 10 May, 2019
- Geoghegan, Tom (April 21, 2013). "BBC News - The story of how the tin can nearly wasn't". Bbc.co.uk. Retrieved June 4, 2013
- Production-scheduling, operations-logistics-supply-chain-terms, business-concepts: mbaskool.com, Retrieved 11 June, 2019
- Choi, Seung-Jin; Burgess (2007). "Practical mathematical model to predict the performance of insulating packages". Packaging Technology and Science. 20 (6): 369–380. Doi:10.1002/pts.762
- Scheduling-9619, production-and-operations-management-tutorial, e-university: wisdomjobs.com, Retrieved 12 July, 2019
- Zabaniotou, A; Kassidi (2003). "Life cycle assessment applied to egg packaging made from polystyrene and recycled paper". Journal of Cleaner Production. 11 (5): 549–559. Doi:10.1016/S0959-6526(02)00076-8
- Y. Schneider; C. Kluge; U. Weiß; H. Rohm (2010). "Packaging Materials and Equipment". In Barry A. Law, A.Y. Tamime (ed.). Technology of Cheesemaking: Second Edition. Wiley-Blackwell. P. 413. ISBN 978-1-4051-8298-0

Casting

CHAPTER 3

The engineering manufacturing process in which materials in a molten state are poured into a mold where they solidify is known as casting. This chapter has been carefully written to provide an easy understanding of the process of casting as well as its diverse types such as centrifugal casting, investment casting and sand casting.

Casting is an engineering manufacturing process generally used for mass production in which materials in a molten state are poured into a mold where they solidify. In this process, complex parts can be manufactured economically and rapidly that otherwise would involve a lot of time if produced by other methods like shaping or cutting. Casting process can be utilized to produce a large variety of parts that are used in different industries, ranging from a small plastic toy to a large gas turbine blade.

Selection of Casting Processes

Different casting processes are available for utilization, and their suitability is influenced by a number of factors including the following:

- Quantity of castings,
- Manufacturing cost,
- Product material,
- Dimensional accuracy required,
- Surface finish required.

CENTRIFUGAL CASTING

Centrifugal casting is a process that delivers components of high material soundness. As a result, it is the technology of choice for applications like jet engine compressor cases, hydro wear rings, many military products, and other high-reliability applications. It has also proven to be a cost-effective means of providing complex shapes with reduced machining requirements and lower manufacturing costs as compared to forgings and fabrications.

In the centrifugal casting process, molten metal is poured into a preheated, spinning die. The die may be oriented either on a vertical or horizontal axis depending on the configuration of the desired part.

By spinning a mold while molten metal is poured into it, centrifugal force acts to distribute the molten metal in the mold at pressures approaching 100 times the force of gravity. The combination of this applied pressure and the engineering mechanics of controlled solidification and secondary refining produces components of superior quality.

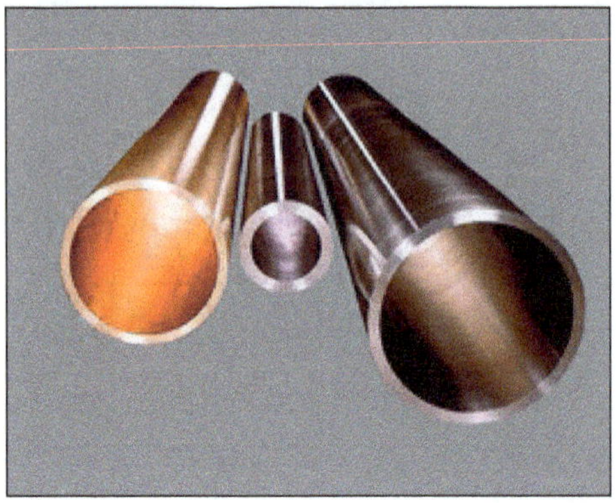

As the die begins to fill, the more dense molten metal is forced to the wall of the spinning die. Directional solidification of sound metal progresses from the O.D. towards the bore, while the less dense material, including impurities, "floats" to the I.D.

Once the casting has solidified, the part is removed from the die and residual impurities in the I.D. are machined away, resulting in a defect-free structure without cavities or gas pockets.

Types of Centrifugal Casting

There are two types of centrifugal casting processes – vertical and horizontal. In addition, some manufacturers offer near-net shaping which combines the benefits of centrifugal casting with O.D shaping, perhaps even with the finished detail of an investment casting.

Vertical Centrifugal Casting

Some manufacturers produce centrifugal components, including some with O.D shaping, in dies rotating about the vertical axis. These vertical castings may achieve that O.D. shaping by inserting graphite, sand, or ceramic molds into the die – resulting in significantly reduced post-processing, like machining or fabrication.

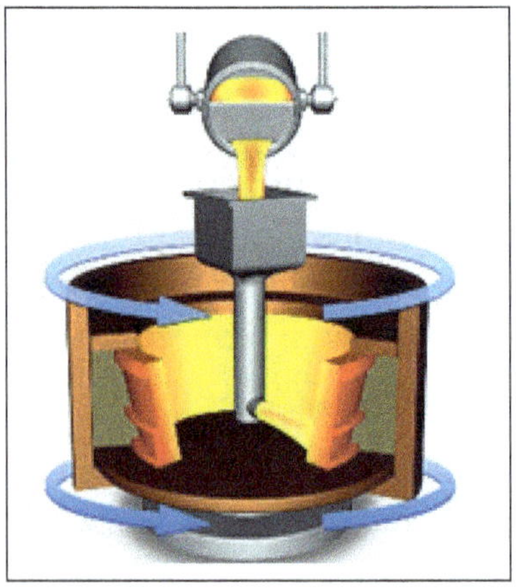

Details on the outside surface of the casting may be modified from the true circular shape by the introduction of flanges or bosses to the inner diameter of the mold. The finished part need not be symmetrical but, in some cases, the casting mold does to maintain balance while spinning.

The inside diameter and therefore the wall thickness of the casting are functions of the amount of metal poured into the rotating mold and the quantity machined away. When casting vertically, the height of the casting will typically be less than twice the width.

Horizontal Centrifugal Casting

Some centrifugal casters produce only horizontal castings where the die rotates about the horizontal axis. This is a cost-effective method for producing high-quality tubular components.

This process is especially suited for long cylindrical parts where the casting length is significantly longer than its outside diameter. This includes straight tube sections, long cylinders with end flanges, or short parts such as rings or flanges where multiple parts can be machined effectively from a straight cylinder.

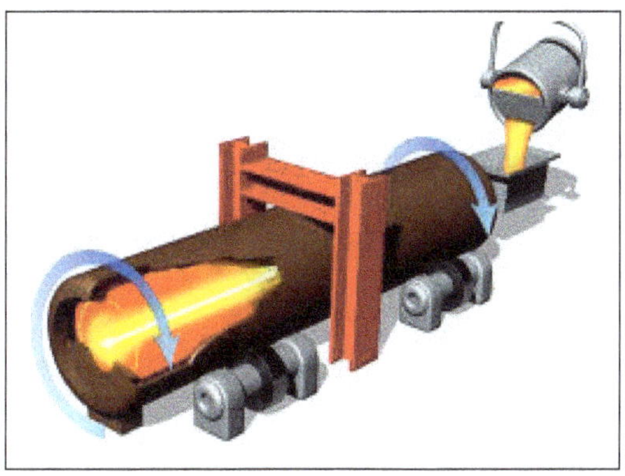

A long steel casting mold is spun at high speed while positioned horizontally. The rotational speed of the mold is high, to offset gravitational forces. Covers are fixed at each end of the mold to contain the molten metal and a pour funnel is used to deliver a specified weight of metal inside the mold.

Just as in vertical casting, the interior dimension of the mold determines the O.D. size of the part, while the amount of metal poured into the mold determines the I.D. size.

Vacuum Centrifugal Casting

Centrifugally casting in a vacuum is used when part detail and control of exposure to the atmosphere is critical since some alloys, including nickel-cobalt super alloys, are reactive to oxygen. In addition to the advantages of casting in a vacuum, the inherent high metal integrity delivered by centrifugal casting is realized, including directional solidification, absence of porosity, and net-shaping. Vacuum centrifugal casting provides products with very high reliability, often used in aerospace and military applications.

Directional Solidification

A key dynamic in the centrifugal process is that solidification is managed to occur in one direction.

The exterior size and shape of the casting is determined by the die (mold) surface, which acts as a chill to initiate solidification. With the aid of exothermal materials that may be added to the I.D. after pouring, along with the heat sink of the mold at the O.D., a temperature gradient results in directional solidification from O.D. to I.D. Liquid metal is continually available to feed the casting, promoting sound metal until the I.D. if finally solidified.

Exterior machining is typically performed to remove surface roughness and maintain concentricity with the bore. However, in some applications, an as-cast O.D. surface is acceptable. Any shrinkage occurring in a centrifugal is at the I.D., which will always be removed by machining.

When to Choose Centrifugal Casting?

- Symmetrical parts that can rotate on an axis.
- Static casting material properties are inadequate.
- Centerline shrink is an issue using other casting processes.
- Limited I.D. features.
- Large parts, up to 135,000 lbs./61,350 kg or more.
- High machining costs for O.D. shaping.
- When required alloy will not work or is not feasible for other processes.
- A uniform grain structure is required.
- Dimensional integrity is required.

CONTINUOUS CASTING

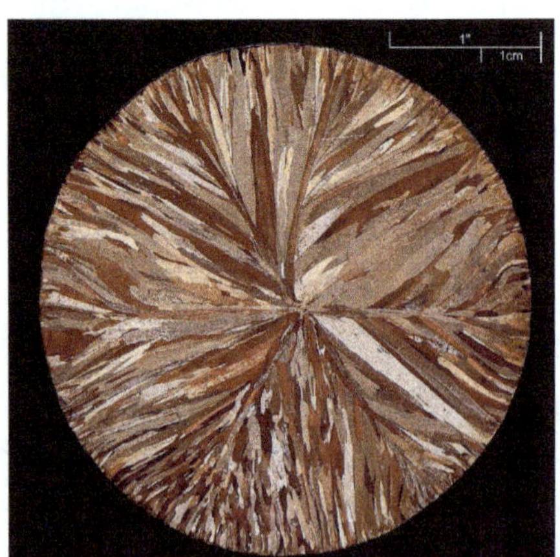

The macrostructure of continuously cast copper (99.95% pure), etched, ∅ ≈ 83 mm.

Continuous casting, also called strand casting, is the process whereby molten metal is solidified into a "semifinished" billet, bloom, or slab for subsequent rolling in the finishing mills. Prior to the introduction of continuous casting in the 1950s, steel was poured into stationary molds to form ingots. Since then, "continuous casting" has evolved to achieve improved yield, quality, productivity and cost efficiency. It allows lower-cost production of metal sections with better quality, due to the inherently lower costs of continuous, standardised production of a product, as well as providing increased control over the process through automation. This process is used most frequently to cast steel (in terms of tonnage cast). Aluminium and copper are also continuously cast.

Sir Henry Bessemer, of Bessemer converter fame, received a patent in 1857 for casting metal between two counter-rotating rollers. The basic outline of this system has recently been implemented today in the casting of steel strip.

Equipment and Process

Low-carbon Steels

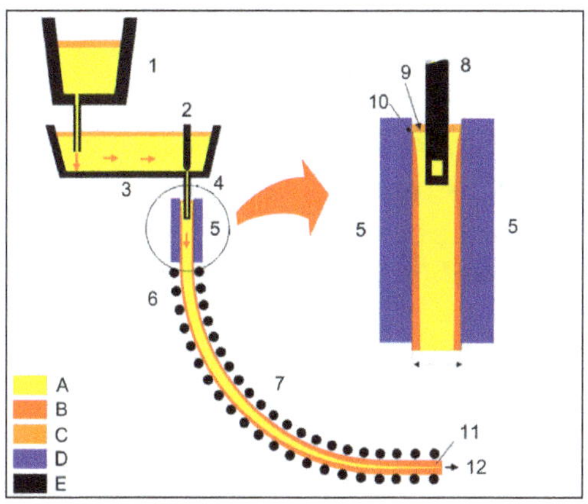

Continuous casting. 1: Ladle. 2: Stopper. 3: Tundish. 4: Shroud. 5: Mold. 6: Roll support. 7: Turning zone. 8: Shroud. 9: Bath level. 10: Meniscus. 11: Withdrawal unit. 12: Slab. A: Liquid metal. B: Solidified metal. C: Slag. D: Water-cooled copper plates. E: Refractory material.

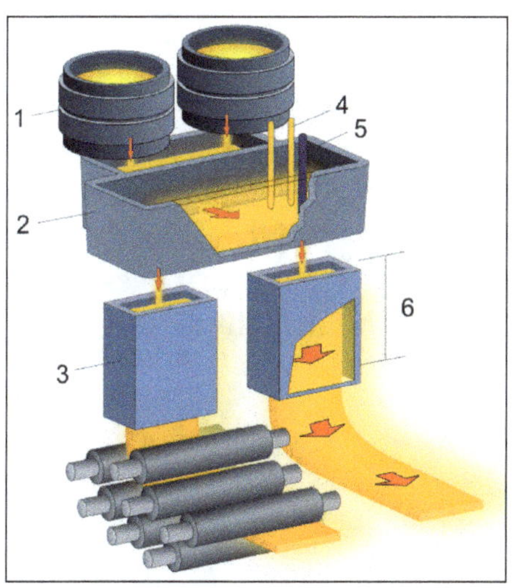

Continuous casting (Tundish and Mold). 1: Ladle. 2: Tundish. 3: Mold. 4: Plasma torch. 5: Stopper. 6: Straight zone.

Molten metal is tapped into the ladle from furnaces. After undergoing any ladle treatments, such as alloying and degassing, and arriving at the correct temperature, the ladle is transported to the top of the casting machine. Usually the ladle sits in a slot on a rotating turret at the casting machine. One ladle is in the 'on-cast' position (feeding the casting machine) while the other is made ready in the 'off-cast' position, and is switched to the casting position when the first ladle is empty.

From the ladle, the hot metal is transferred via a refractory shroud (pipe) to a holding bath called a tundish. The tundish allows a reservoir of metal to feed the casting machine while ladles are switched, thus acting as a buffer of hot metal, as well as smoothing out flow, regulating metal feed to the molds and cleaning the metal.

Metal is drained from the tundish through another shroud into the top of an open-base copper mold. The depth of the mold can range from 0.5 to 2 metres (20 to 79 in), depending on the casting speed and section size. The mold is water-cooled to solidify the hot metal directly in contact with it; this is the *primary cooling* process. It also oscillates vertically (or in a near vertical curved path) to prevent the metal sticking to the mold walls. A lubricant (either powders that melt on contact with the metal, or liquids) is added to the metal in the mold to prevent sticking, and to trap any slag particles—including oxide particles or scale—that may be present in the metal and bring them to the top of the pool to form a floating layer of slag. The shroud is set so the hot metal exits it below the surface of the slag layer in the mold and is thus called a submerged entry nozzle (SEN). In some cases, shrouds may not be used between tundish and mold ('open-pour' casting); in this case, interchangeable metering nozzles in the base of the tundish direct the metal into the moulds. Some continuous casting layouts feed several molds from the same tundish.

In the mold, a thin shell of metal next to the mold walls solidifies before the middle section, now called a strand, exits the base of the mold into a spray chamber. The bulk of metal within the walls of the strand is still molten. The strand is immediately supported by closely spaced, water-cooled rollers which support the walls of the strand against the ferrostatic pressure (compare hydrostatic pressure) of the still-solidifying liquid within the strand. To increase the rate of solidification, the strand is sprayed with large amounts of water as it passes through the spray-chamber; this is the *secondary cooling* process. Final solidification of the strand may take place after the strand has exited the spray-chamber.

It is here that the design of continuous casting machines may vary. This describes a 'curved apron' casting machine; vertical configurations are also used. In a curved apron casting machine, the strand exits the mold vertically (or on a near vertical curved path) and as it travels through the spray-chamber, the rollers gradually curve the strand towards the horizontal. In a vertical casting machine, the strand stays vertical as it passes through the spray-chamber. Molds in a curved apron casting machine can be straight or curved, depending on the basic design of the machine.

In a true horizontal casting machine, the mold axis is horizontal and the flow of steel is horizontal from liquid to thin shell to solid (no bending). In this type of machine, either strand or mold oscillation is used to prevent sticking in the mold.

After exiting the spray-chamber, the strand passes through straightening rolls (if cast on other than a vertical machine) and withdrawal rolls. There may be a hot rolling stand after withdrawal to take advantage of the metal's hot condition to pre-shape the final strand. Finally, the strand is cut into predetermined lengths by mechanical shears or by travelling oxyacetylene torches, is marked for identification, and is taken either to a stockpile or to the next forming process.

In many cases the strand may continue through additional rollers and other mechanisms which may flatten, roll or extrude the metal into its final shape.

Casting Machines for Aluminium and Copper

Molten aluminum pours into this casting die (top view of die).

The resulting Aluminum blanks (after cutting to size).

Aluminium and copper can be cast horizontally and can be more easily cast into near net shape, especially strip, due to their lower melting temperatures.

Range of Continuously Cast Sections

- Casting machines are designated to be billet, bloom or slab casters.
- Slab casters tend to cast sections that are much wider than thick:
    - Conventional slabs lie in the range 100–1600 mm wide by 180–250 mm thick and up to 12 m long with conventional casting speeds of up to 1.4 m/minute; however slab widths and casting speeds are currently increasing.
    - Wider slabs are available up to 3250×150 mm.
    - Thick slabs are available up to 2200×450 mm at a specific steel facility, generically ranging typically from 200mm to 300mm.
    - Thin slabs (low-carbon steel): 1680×50 mm at a specific facility, generically ranging from 40mm to 110mm thick depending on an individual machine's design.

- Conventional bloom casters cast sections above 200×200 mm. The bloom length can vary from 4 to 10 m.
- Billet casters cast smaller section sizes, such as below 200 mm square, with lengths up to 12 m long. Cast speeds can reach up to 4 m/minute.
- Rounds: either 500 mm or 140 mm in diameter.
- Conventional beam blanks: look similar to I-beams in cross-section; 1048×450 mm or 438×381 mm overall.
- Near net shape beam blanks: 850×250 mm overall.
- Strip: 2–5 mm thick by 760–1330 mm wide.

Startup and Control of the Process

Starting a continuous casting machine involves placing a dummy bar (essentially a curved metal beam) up through the spray chamber to close off the base of the mould. Metal is poured into the mould and withdrawn with the dummy bar once it solidifies. It is extremely important that the metal supply afterwards be guaranteed to avoid unnecessary shutdowns and restarts, known as 'turnarounds'. Each time the caster stops and restarts, a new tundish is required, as any uncast metal in the tundish cannot be drained and instead freezes into a 'skull'. Avoiding turnarounds requires the meltshop, including ladle furnaces (if any) to keep tight control on the temperature of the metal, which can vary dramatically with alloying additions, slag cover and deslagging, and the preheating of the ladle before it accepts metal, among other parameters. However, the cast rate may be lowered by reducing the amount of metal in the tundish (although this can increase wear on the tundish), or if the caster has multiple strands, one or more strands may be shut down to accommodate upstream delays. Turnarounds may be scheduled into a production sequence if the tundish temperature becomes too high after a certain number of heats or the service lifetime of a non-replaceable component (i.e., the submerged entry nozzle (SEN) in a thin-slab casting machine) is reached.

Many continuous casting operations are now fully computer-controlled. Several electromagnetic, thermal, or radiation sensors at the ladle shroud, tundish and mould sense the metal level or weight, flow rate and temperature of the hot metal, and the programmable logic controller (PLC) can set the rate of strand withdrawal via speed control of the withdrawal rolls. The flow of metal into the moulds can be controlled via three methods:

- By stopper rods that descend through the tundish,
- By slide gates at the top of the mould shrouds,
- If the metal is open-poured, then the metal flow into the moulds is controlled solely by the internal diameter of the metering nozzles. These nozzles are usually interchangeable.

Overall casting speed can be adjusted by altering the amount of metal in the tundish, via the ladle slide gate. The PLC can also set the mould oscillation rate and the rate of mould powder feed, as well as the flow of water in the cooling sprays within the strand. Computer control also allows vital casting data to be transmitted to other manufacturing centres (particularly the steelmaking furnaces), allowing their work rates to be adjusted to avoid 'overflow' or 'underrun' of product.

Problems

Contamination by Oxygen

While the large amount of automation helps produce castings with no shrinkage and little segregation, continuous casting is of no use if the metal is not clean beforehand, or becomes 'dirty' during the casting process. One of the main methods through which hot metal may become dirty is by oxidation, which occurs rapidly at molten metal temperatures (up to 1700 °C for steel); inclusions of gas, slag or undissolved alloys may also be present. To prevent oxidation, the metal is isolated from the atmosphere as much as possible. To achieve this, exposed liquid metal surfaces are covered – by the shrouds, or in the case of the ladle, tundish and mould, by synthetic slag. In the tundish, any inclusions that are less dense than the liquid metal – gas bubbles, other slag or oxides, or undissolved alloys – may also float to the surface and be trapped in the slag layer. While the tundish and mold fill for the first time at the start of a casting run, the liquid is badly contaminated with oxygen and the first items produced are typically quarantined or diverted to customers who do not require top-quality material.

Breakouts

A major problem that may occur in continuous casting is *breakout* of the liquid metal: for whatever reason, the solid shell of the strand breaks and allows the still-molten metal contained within to spill out and foul the machine. In most industrial environments this event is very costly as it leads to a shutdown of the strand and typically requires an extended turnaround involving removal of the spilled material from within the strand equipment and/or replacement of damaged machinery. A breakout is usually due to the shell wall being too thin to support the liquid column above it, a condition which has several root causes often related to heat management. Improper cooling water flow to the mould or the strand cooling sprays may lead to inadequate heat removal from the solidifying metal, causing the solid shell to thicken too slowly. If the metal withdrawal rate is too fast, the shell may not have time to solidify to the required thickness even with enhanced cooling sprays. Similarly, the incoming liquid metal may be too hot and the final solidification may occur further down the strand at a later point than expected; if this point is below the straightening rolls, the shell may break from stresses applied during straightening.

A breakout can also occur as a result of physical irregularities or damage to the shell occurring within the mould during the initial seconds of solidification. Excessive turbulence within the mold may cause an irregular shell pattern that grows abnormally or it may entrap slag droplets within the shell which reduces the wall strength. A common occurrence is for the shell to stick to the mould's surface and tear; modern instrumented molds and computer control systems typically detect this and slow the caster down temporarily to let the wall refreeze and heal while it is still supported in the mould. Should the tear occur near the exit of the mould or be of unexpected severity, the shell may still fail in a breakout once it emerges from the mould wall. If the incoming metal is severely overheated, it may be preferable to stop the caster than to risk a breakout. Additionally, lead contamination of the metal (caused by counterweights or lead-acid batteries in the initial steel charge) can form a thin film between the mould wall and the steel, inhibiting heat removal and shell growth and increasing the risk of breakouts.

Other Considerations

Another problem that may occur is a *carbon boil* – oxygen dissolved in the steel reacts with also-present carbon to generate bubbles of carbon monoxide. As the term *boil* suggests, this reaction is extremely fast and violent, generating large amounts of hot gas, and is especially dangerous if it occurs in the confined spaces of a casting machine. Oxygen can be removed by "killing" it through the addition of silicon or aluminium to the steel, which reacts to form silicon oxide (silica) or aluminium oxide (alumina). However, too much alumina in the steel will clog the casting nozzles and cause the steel to 'choke off'.

Computational fluid dynamics and other fluid flow techniques are being used extensively in the design of new continuous casting operations, especially in the tundish, to ensure that inclusions and turbulence are removed from the hot metal, yet ensure that all the metal reaches the mould before it cools too much. Slight adjustments to the flow conditions within the tundish or the mould can mean the difference between high and low rejection rates of the product.

Starter Bar

The starter bar, also called a dummy bar, has a free end portion which is flexible for storage and a substantially rigid portion at the end which plugs the mold. The starter bar is constructed in discrete blocks secured to one side of a planar spine provided in segments and arranged end to end. Adjustable spacers in the form of tapered blocks are disposed between the blocks of the bar to allow the starter bar to be self-supporting in a curved configuration corresponding to the casting path. A more flexible spine in the end portion of the starter bar allows the starter bar to be curved to a tighter radius than that of the casting path while the blocks fan out in an unsupported configuration. A storage ramp is provided to support the flexible end in the stored position. Before a cast is started, the starter bars are fed through the caster (in reverse direction of casting) using hydraulic actuators. Once fed all the way to the bottom of the mold, the process of packing the mold can continue to ensure a smooth start up.

Direct Strip Casting

Direct strip casting is a continuous casting process for producing metallic sheet directly from the molten state that minimizes the need for substantial secondary processing. For low-carbon sheet steels, this is a relatively new process which has only achieved commercial industrialization within the last decade.

Twin-belt Continuous Casting

Twin-belt continuous casting is a continuous casting process that produces high volume continuous metal bar or strip of constant rectangular cross section. Twin-belt continuous casting employs a moving mold consisting of parallel carbon-steel belts held in tension as top and bottom casting surfaces. Chains of rectangular steel or copper blocks moving with the belts and spaced according to the desired cast width form the sides of the mold.

Molten metal is introduced into the twin-belt continuous casting machine from a tundish through a nozzle placed between the casting belts. The metal is cooled by direct contact with the belts which are in turn cooled by high pressure recirculating water. Various coatings can be applied to the belt casting surfaces to provide required mold interface characteristics and prevent adhesion.

The cast metal from the twin-belt continuous casting machine is synchronized with, and directly fed into, a hot rolling mill. Combining the casting and rolling operations can result in significant energy and cost savings over other casting processes which incorporate intermediate cast and reheat steps.

1. Metals cast on twin-belt continuous casting machines: Copper (Bar, Strip, Anode), Aluminum(Strip), Zinc (Strip), Lead (Strip).

2. Production rates and speeds: Twin-belt continuous casting rates range up to 60 tons per hour at speeds up to 14 meters per minute.

Twin-belt continuous casting is a near net shape casting process, which significantly reduces the need for secondary rolling or forming operations. For example, when casting copper anode plate the cast slab is not rolled but rather sheared directly into distinct anode plates.

The cooling belts are typically made of low carbon steel and are held under tension within the casting machine to ensure flatness and accuracy. As a "cold" belt enters the mold region, it is heated in the cast zone and is subject to powerful forces caused by thermal expansion. When casting wide strip, these forces must be controlled to eliminate buckling and reduce thermal distortion of the belt at the mold entrance. These forces can be controlled by preheating the belts before mold entry, or by magnetically stabilizing them once they have entered the mold.

1. Belt preheating: For wide strip casting, a belt preheating system can be used to bring the belt up to 150 °C or higher immediately prior to entering the casting mold, reducing the effects of cold framing. Induction heating coils can be used across the width to preheat each belt. In addition to preventing thermal distortion, the high preheat temperature serves to eliminate any moisture present on the belt surface.

2. Magnetic stabilization: When casting wide strip, the tendency of localized thermal distortion can be resisted by the use of high-strength, magnetic belt back-up support rolls within the mold region. The moving belt is held against the support rolls by magnetized rotating fins maintaining the belt in a flat plane.

Within the twin-belt continuous casting machine, molten metal progressively solidifies on the mold surfaces as it moves through the mold region, with a sump of molten metal present between the solidifying outer surfaces. Belt coatings, texture, and gas layer modifications are used to fine tune the heat transfer rate from the cast metal to the belt. Full thickness solidification can occur as early as 30% of the way through the mold for thin strip, or up to 2 m beyond the mold exit for large bar where exit water spray cooling and roller support are required.

3. Closed pool feeding: When casting certain metals such as aluminum, a fully closed pool "injection" metal feeding system can be employed. Here, the metal is introduced under slight pressure into the closed mold cavity. Metal flow is controlled by maintaining a preset level in the tundish. The feed snout, or nozzle, is typically made from a ceramic material which is thermally stable and permeable to gases being released from the flowing metal.

4. Open pool feeding: When casting other metals, such as copper, zinc and lead, an open pool feeding system is often used. In this case, the upper belt pulley is offset downstream from the bottom pulley. Metal flows through an open trough or tundish into a standing pool of molten metal formed at the convergence of the belts. Shrouding gases may be employed to protect against oxidation.

5. Mold tapering: The twin-belt casting machine differs from other moving mold casting machines in that all four mold surfaces are independent. This allows the mold surfaces to be tapered to remain in contact with the cast product as it shrinks. The high velocity cooling water, which is continuously applied to the backside of the belt, impinges on the belt and creates a force on the belt. This force acts to press the belt against the surface of the strip or slab as it shrinks, keeping the belt in close contact with the cast product throughout the mold. Each side of the mold is formed by an endless chain of dam blocks,which are held against the cast strip by adjustable spring-loaded guides.

6. Molten metal level control: To accommodate high casting speeds and maintain as high a pool level as possible, non-contact electromagnetic metal level indicators can be used to sense the pool level in the casting machine.

7. Aluminum or copper strip casting: Commercial twin-belt continuous strip casting machines are capable of producing as-cast dimensions from 10–35 mm thick, and up to 20–35 mm wide. After being directly fed into a hot rolling mill, the as-cast strip is typically rolled down to 1–3 mm thickness strip.

8. Copper bar casting: As-cast dimensions range from 35–75 mm thick, and from 50-150 mm wide. After being directly fed into a hot rolling mill, the as-cast bar is typically rolled into 8 mm diameter rod to be used for wire drawing.

9. Copper anode casting: Special dam blocks which contain anode lug molds and a traveling hydraulic shear are added to the twin-belt casting machine to continuously cast net shape copper anodes. Anode width of approximately 1 meter (excluding lugs) and thicknesses from 16 mm to 45 mm. The primary advantage of this process is uniformity of the as-cast anode in terms of size and surface quality. Anodes cast using this process do not require additional preparation after casting.

10. Mold length: The mold length ranges from approximately 2000 mm for strip casting machines and up to 3700 mm for copper bar casting machines.

EVAPORATIVE PATTERN CASTING

Evaporative pattern casting (EPC) Process is a sand molding process that makes use of evaporative patterns produced industrially by steam molding or machined out from a block of expandable polystyrene foam (EPS). The pattern is buried in the sand mold and the melted casting material is poured into the mold without removing pattern, unlike the traditional sand casting method that makes use of wooden, plastic and metallic pattern which is removed from the mold before molten metal is advanced.

It was in 1956 that Shroyer made a documentation of his work on EPC where green sand was used as the molding material. The author machined a shape from expanded polystyrene (EPS) and supported it inside a flask with bounded sand. Another researcher called Flemmings came up in 1964 to use unbounded silica grains for the process. Foundry men later came to differentiate between the two; calling the first the full mold and the second the lost foam processes. The subtle difference only lied in the use of bounded (green) sand and loose sand.

The process became known to foundry men in 1958 and many trade names have been formed to describe it. They include styrecast, replicast, full mold, lost foam and evaporative process. A distinguishing feature of this type of sand process is that an evaporative pattern made of polystyrene foam is used. It is coated with a coating developed from a refractory material usually silica or zircon sand before being buried in the mold. It is not removed from the mold once it is buried.

After 1964, some works have been done by researchers to understand the variables that affect the process. Some works in the automotive industry have also been done. General motors have produced a number of automotive components with the method. This started in the year 1980 and they are still in the use of the process. The full mold process unlike the conventional sand casting process uses patterns and pouring system, which retains the pattern in the mold, thus justifying the title full mold. The term 'evaporative' stands as a family name for all the trade names that have so far been coined out for evaporative pattern casting processes. This is because the pattern evaporates when there is contact between it and liquid metal; all the features are replicated.

By evaluation and practices, the process is known to take care of intricate and complex shapes without the use of cores; this is not achievable in traditional sand casting methods. The surface finish of the castings produced has an edge over the ones of green sand method. Very slender elements are produced with ease using EPC which is very difficult in other sand casting methods. Precision in terms of dimensional accuracy is better.

With the trend in the automotive industries; where engine blocks and cylinder heads are made with aluminum alloy, the evaporative pattern casting process has found a place for effective production. This has opened up endless research opportunities aimed to improve on the use of metal alloy components obtained by EPC.

Patterns used in EPC process are produced locally by machining them from blocks of polystyrene material and industrially by steam molding of the beads. The ones produced industrially are for mass production. The commonly used pattern material is expandable polystyrene (EPS). Polyalkylene carbonate (PAC) is used to manufacture ferrous castings. Another material that is used for EPC is the poly methyl methacrylate (PMMA). Metal tooling with techniques that have been developed for industrial packaging is used. Precision machine tools are required for EPC process. The tooling material is aluminum alloy. It is used because of its exceptional properties which are not found in other alloys. The properties include good thermal conductivity, good erosion resistance, fair strength, light weight and good surface finish upon machining.

The process of producing pattern material is regarded as polymerization. In local development of EPC Process, the patterns are cut manually using knives, blades, pre-heated wire etc. and then smoothened out by emerald paper. At times when parts are to be joined together, the use of glue is employed. Juice from rubber plant can as well be used to join patterns. After the production of the pattern and it is assembled with the gating system, it is coated with refractory material which may be wet (slurry) or dry. The coating provides a barrier between the flowing motel metal and the molding sand to prevent metal penetration. Other benefits are offered. An example is the quality of casting surface produced.

The gating system assembled with the pattern is informed by the weight of the casting to be poured. Immediately the pattern that is coated is dry, it is taken and buried in the mold. Care is taken so that the coating layer is not peeled off. Molding sand is packed into the molding flask and rammed carefully so that the pattern is not distorted and the mold is poured. After solidification, shake out process follows and cleaning of the casting is done. The molding sand is entirely reclaimable. The process offers many benefits to both the casting producers and users in terms of time reduction and low cost in the production of the patterns and low cost of purchase of pattern materials.

Mold production entails the placement of the coated pattern assembly in green sand and rammed to obtain compaction of the grains. In the full mold, green sand is used as molding material. Things required in mold production are: molding sand, molding boxes, pattern and vibration. Following vibration, the mold is prepared for casting, which may simply involve the location of a pouring basin above the sprue entrance. However, to prevent fluidization/erosion of the sand during pouring, caused by the evolved gaseous pattern decomposition products, mold weights may be necessary, or the top of the box may be covered with a plastic sheet and a soft vacuum applied.

When pouring the liquid metal into prepared molds, it must flow full right from the pouring cup on top of the sprue in the gating system down to the mold cavity. This offers advantages in that the mold would not collapse, the sprue evaporates completely, air aspiration is avoided and sound casting is obtained. The pouring rate is determined by the density of the foam, permeability of the sand and casting material. Gating design is therefore done carefully to achieve sound castings. The elements of gating system include pouring cup, down sprue, sprue base well, runners, ingates and risers. Careful design of each of these elements brings about soundness in castings. The gating ratio is informed by the type of material to be cast. For aluminum alloys, non-pressured gating ratios like 1:2:4, 1:3:3 or 1:4:4 is used. This refers to the cross-sectional areas of the choke, runners and ingates. For the ferrous materials, pressurized gating ratios are used like 4:3:1, 2:1:8, 1:1:4 and 1:2:1.

Once liquid metal is poured in EPC mold both physical and chemical reactions start to occur. EPC is known to be influenced by many variables. The combination of these variables determines whether sound casting would be obtained. The process begins by mold filling with the liquid metal, followed by transfer of heat energy from the liquid metal to the evaporative pattern in the mold. The flow of liquid metal to make contact with the pattern is also observed and when it eventually makes contact, decomposition of the foam begins, leading to pyrolysis products. Pressure in the mold is built up as a result of these products and reactions. The pressure determines what quality of the casting that would be obtained.

As a unique technique to produce integrated components which otherwise must be cast in several parts, it has drawn great interest from both academia and industry. Efforts have been made over the years to understand variables and physico-chemical activities that significantly affect the process and initiate means of monitoring and controlling them. These research efforts have enabled rapid growth of EPC production around the world. This has cumulated to the increase of the products of the process in the automotive and similar industries. Researches are still on-going to maximize the benefits.

It has been observed that the EPC offers many advantages over the traditional/conventional casting processes. One of the advantages is that it eliminates the requirements of cores for internal structures because foam patterns which are produced by steam molding in industrial set-up or machined from a block of polystyrene are shaped desirably to bring out the casting. This machining method also allows several parts to be assembled and makes complex casting designs possible. Other advantages include increased dimensional accuracy, saved material and costs as castings produced with the process require minimal fettling.

In EPC Process the molding sand is reusable because no binders are used, which makes EPC a more environmentally friendly process. This process also eliminates issues like dewaxing and mold firing that occurs in the lost wax process. It has been shown that the mechanical properties of lost foam castings and full-mold castings have cutting edge over those of other sand castings. Other attractive features include near-net-shape casting, elimination of parting lines and excellent surface finish. Table shows a comparison among the various casting processes on labor cost, equipment and surface finish. Studies show that all ferrous and nonferrous metals have been successfully cast using EPC process.

Table: Summary of casting processes.

Type of process	Labor cost per unit	Equipment cost	Surface finish μ m CLA	Accuracy mm	Minimum section mm
Sand (green)	Medium	Low	500–1000	±2.5	5.0
Shell	Low	Medium	100–300	±0.25	2.5
Centrifugal	Low	Medium	100–500	±0.7	8.0
Investment	High	Medium	25–125	±0.06–25	0.6
Die casting gravity	Low	Medium	100–250	±0.4 + 0.05 per 25 mm	2.5
Die casting low press	Low	Medium	40–100	±0.05 + 0.05 per 25 mm	1.2
continuous	Low	High	100–200	±0.12 per 25 mm	8.0
EPC	Low	Low	60–300	±0.05 + 0.05 per 25 mm	1.0

One outstanding advantage that EPC process offers is that there is minimal number of rejected castings. It could then be inferred that there is a saving on cost of purchase of furnace fuel used in firing and melting and energy to remelt the scraps or rejected products is reduced. The casting material is also optimized as it does not end in scraps; sound castings are produced with them.

In overall evaluation EPC should be preferred to the traditional casting methods. Therefore in where energy supply is not adequate the foundry engineer could make use of the EPC for these afore mentioned benefits and the rapid production of the patterns with less energy expended unlike the patterns used in the conventional methods where large amount of energy is expended.

With the advantages and benefits that the process offers, not all foundries are known to embrace it. However, predictions have been made that soon very many foundries would use the process; looking at it from the perspective of automotive industries. It is believed that if the process parameters that affect the EPC process are properly monitored and controlled and there is proper gating system design and melting and pouring practices, defect free castings should be obtained.

In order to obtain castings with good surface finish free from defects researchers both in industries and academia are giving attention to both experimental, modeling and theoretical studies of the variables known to influence the process. It has been recognized that the pyrolysis products are the main source of defects in EPC.

The EPC process is very sensitive and known to be complex because of the very many variables that affect it. Experiments and research efforts have not been able to adequately quantify certain parameters like the heat transfer coefficient existing between the liquid metal flow and the decomposing foam. Therefore a range of the coefficient has been proposed to be 40–160 W/m^2 K.

According to Mohammed computational models and simulations have been developed to further understand the interactions of various variables in EPC process in order to help reduce defect formation in the castings. Since the heat transfer from the molten metal to foam pattern plays a very important role, models without sufficient consideration of the foam pyrolysis process do not capture the effect of process parameters on the defect formation in the castings.

Evaporative Pattern Casting Process Steps

All the activities are diagrammatically illustrated in figure. It is of interest to see the steps simply outlined but every step needs to be carefully monitored and controlled because of the physico-chemical interactions in the steps coupled with the process parameters that are involved for the purpose of producing consistent and high quality castings.

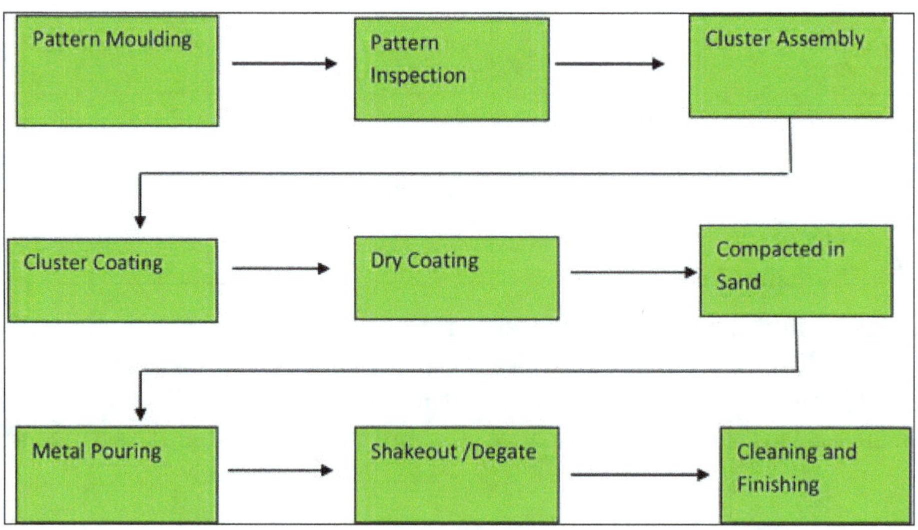

Steps of EPC process.

The very many variables of EPC process grouped into six categories have made the process to react to changes that occur in the system. The combined and individual effects of these variables sometimes produce significant effects on the process and its products. The chain of systems observed to make castings by EPC process are divided into two main classes:

- Evaporative pattern production, assembly and inspection.
- Casting production and inspection.

The first division is a six-step operation. The step 1 is the pattern molding. This is done industrially by injection process. Beads of polystyrene foam are injected into molding machine under pressure. The foam material together with the amount of applied pressure for compaction determines the density of the polystyrene pattern that would be employed for casting. This is an industrial option. On the other hand, patterns are machined from a block of polystyrene foam with a device called heated wire. Step 2 is the inspection of the patterns produced industrially or manually. Dimensional accuracy and correctness of the shape are inspected at this level. Distortion of the patterns is avoided. This is paramount to the EPC Process. Many a time, EPC Process involves casting of intricate shapes such that a shape will be broken into segments and patterns will be produced for these segments. The patterns of the segments are joined together with glue. The joining process is referred to as cluster assembly which is step 3. To provide ability to withstand thermal energy from advancing liquid metal and prevent sand penetration in evaporative pattern castings, refractory coating which is made from highly refractory materials is sprayed on the patterns. Step 4 takes care of this. The patterns are left to dry. The drying period is within a matter of seconds; maximum 15 s. This is possible because of the reducer/carrier which is usually methyl alcohol at 99% concentration that is employed in the transfer of the coating to spraying the pattern. The drying process is done in step 5. Step 6 deals with compaction of the coated patterns in the sand mold. The sand mold is prepared with molding sand usually river sand whose grain fineness number has been determined and some other properties which molding materials should have like green strength, refractoriness, hot strength and so on. Immediately the bed is laid, the pattern is put inside the molding flask then it is filled with molding and backing sand. The pattern is not removed from the mold.

The second division deals with casting production and inspection. Three steps are involved: Step 1 deals with metal pouring. Casting materials are charged into the furnace and allowed to melt. Each metal has its melting temperature. The foundry man takes the melt from the furnace with pouring equipment and pours into the prepared mold. Solidification of the casting takes place inside the mold; Step 2 takes care of shakeout and de-gating of the casting. The process of removing the casting from the mold is the shakeout while removal of sprue, runner bar and ingates from the casting is the de-gating; Step 3 which is the last of all sees to the cleaning and finishing of the casting. Here brushes are employed to clean by removing sand attached to the casting. Unnecessary projections are also cut off.

Advantages and Disadvantages of EPC Process

From an environmental and economic perspective, one of the principal advantages of EPC Process is the fact that the casting sand can be reused. This is not so with the traditional green sand casting method as it will require chemical treatment to maintain the properties of the molding sand with will have negative impact of the environment when the sand is finally disposed. The process offers manufacture of complex shapes without the need for cores which must be provided for in production of holes and passages when green sand mold method is used. EPC Process does not present lines that separate molds into several parts which are called parting lines. Based on these afore, researchers have sub mitted that the process reduces machining cost and energy use.

Even with the advantages that researches have outlined which include good surface finish, high dimensional accuracy and reduction in cost of production which make the EPC process to be competitive; it is not without its disadvantages. It is true that the process offers good surface finish but it is subject to the surface of the pattern. The pattern density has significant effect on the quality of casting produced with it. The density values of the patterns vary due to level of compaction of the beads by steam molding. The foams, if not carefully handled may be distorted or permanently damaged. On the whole the disadvantages recorded in the process are associated with the EPS.

Problems and Challenges in EPC Process

EPC Process is known to offer many advantages over the traditional sand molding processes like rapid production of patterns and at low costs. However, there are pronounced disadvantages and problems that have been identified. These problems are known to occur because of the sensitivity of the Process to the variables. The chemistry of decomposition of the polystyrene foam which is used as the pattern material is vital in the overall success. The pyrolysis products therefore must escape from the molds as fast as possible to have castings free of defects.

Some problems are associated with the process majorly referred to as casting defects. The EPS foam pattern undergoes a series of complex reactions: collapse, liquefaction, vaporization and depolymerization. The pyrolysis liquid and gaseous products are potential defect sources like blow holes and the effective elimination of these degradation products is important to produce quality castings. Common EPC defects such as internal porosity, folds and surface defects are all pyrolysis product related.

Variable like refractory coating, pattern density uniformity, and pouring temperature are critical factors affecting formation of defects in the castings. Porosity defects are created when a fast moving metal front engulfs portions of the foam pattern which form voids in the solidified castings. Folds are caused when two streams of molten metal meet and pyrolysis products at the metal front prevent the two streams of metal from fusing. Surface defects are present at the surface of the casting, and are a result of foam pyrolysis products trapped at the metal-coating interface.

Applications of Evaporative Pattern Casting Process

The application of the EPC process is dynamic. It has been applied in the production of aluminum and its alloys. Iron founding also makes use of the process for production of various products. Magnesium and it alloys are now being cast using the method. The first public recognition of the application of the process was in 1980 when General Motors produced some automobile components with it. Aluminum railway valve body, aluminum cylinder head, pipe fittings and shaft hubs are now being produced by foundries in Canada, Italy, USA and Netherlands.

EPC process Parameters

EPC Process is relatively new and from works of researchers, many process parameters have been known to affect/influence the process. It is also believed to be very sensitive as these variables affect the soundness of the castings produced by it. In both local and industrial applications of the process, difficulties have been encountered. The difficulties have led to casting defects. Some of the process parameters are the patterns, pattern coating, pouring temperature, vibration, gating ratios and geometry of components.

Patterns in EPC Process

Patterns are made of expanded polystyrene foam; poly methyl methacrylate and polyalkylene carbonate. The materials are light in weight. They must be carefully handled after production to avoid distortion. The expanded polystyrene is used for aluminum and its alloys; it is not used in iron founding because of the defects that it causes for the casting. It is known that the materials have different density values and this has effects on the quality of castings produced. It is then required that there must be consistency in the material used as pattern for consistent quality in terms of the mechanical properties and the microstructures of the castings. It is observed that the pattern used in EPC is in sharp contrast to the ones employed in the traditional sand casting methods where wooden, metallic and plastic materials are used to create cavity in the molds and then removed before pouring casting material into the molds.

Pattern Properties

The properties of the pattern significantly affect the casting quality. To a very large extent, defects caused in castings are attributed to the non-uniformity in pattern density. Consistency should be observed in the properties of the pattern, especially, the density which is a result of the bead compaction. The density affects the flow of molten metal expected to displace the pattern material buried in the sand mold when the compaction of the beads of one side is different from the other side. Instead of having uniform flow of molten metal, the metal tends to flow to low density areas, thus causing folds and foam inclusion in the casting because metal flow faster in the area of low density showing low bead compaction. The decomposition of the patterns may be by scission or unzipping; a number of reactions are involved in this before completion and patterns rejected usually become environmental problems.

Pattern Coating

Pattern coating which is a mixture of refractory material and binder or many a times only the refractory material, when applied on patterns, forms a solid layer on the pattern when it is dry. The coating must be permeable to allow gas to escape into the surrounding; otherwise the gas would be trapped, causing defects in the castings. Houzeaux and Codina showed that the escape of the gas from the mold is subject to coating applied on the foam. Chen and Penumadu posited that mold filling times decreased with permeability of the coatings.

Depending on the type of coating, the time to fill the mold with the liquid metal is affected. The gas generated must escape continuously from the mold to avoid defects. The pyrolysis products are the main causes of defects in EPC Process. The release of the gas products should be in a timely manner; if they are too fast, it leads to mold collapse because of pressure drops such that the coating layer is no longer supported and could not therefore bear the weight of the sand. An ideal pattern coating must allow gaseous and liquid foam degradation products to be transported out of the casting in a timely and balanced manner. Variables such as coating material, percent solid, viscosity, liquid absorption capability, coating thickness and gas permeability affect the quality of casting".

The application of the pattern coating must be carefully done on the pattern to ensure that it is consistent. If the coating is applied wet, the wettability must be consistent. If it is applied dry, the thickness must be consistent. This will allow for controlled liquid absorption and gas permeability. The coating layer provides clean and smooth surfaces on the castings because it is easily removed.

As regards the particle size of the refractory coating, 200 meshes (75 μm) are recommended. However, the particle size can be reduced to 65 μm. This gives good surface finish to the castings produced with it. The surface roughness obtained in the castings when the coatings are used is subject to the type of casting method employed.

Usually one of the three methods; spraying, dipping, swabbing, is used to apply the refractory coating to the surface of patterns. Spraying has proved to be the most effective in that uniform spread of the coating is obtained leading to almost uniform thickness. In reality constant thickness cannot be obtained for all patterns. The dipping process will not give uniform spread of the coating. There is great variation in the thickness that would be obtained. In the swabbing, the brush used in the painting makes marks on the pattern thereby reproducing the marks on the castings produced with it.

Structure and Thermal Degradation of Foam Pattern

Polystyrene was discovered in Germany in 1839 by Edward Simon and is a derivative of petroleum produced by polymerization of monomer styrene. The chemical structure of the polystyrene contains only carbon and hydrogen atoms with benzene ring attached to it and it is then classified as hydrocarbon. The chemical formula is $(C_8H_8)_n$. The polymerization process of producing the foam is illustrated in the Figure. In EPC Process, thermal degradation of the polystyrene occurs at elevated temperature.

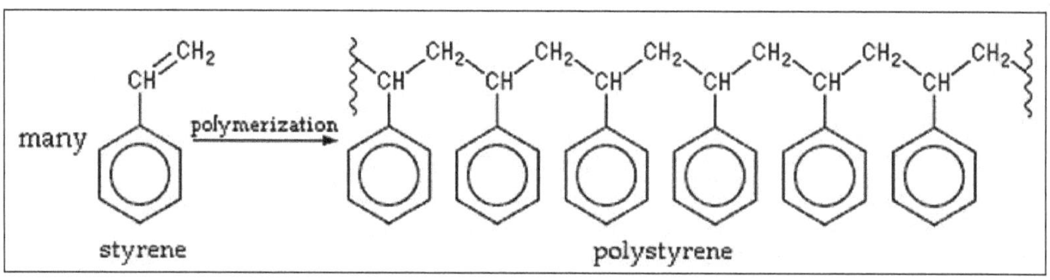

Polymerization process.

In the physical model as a result of heat energy transferred from the liquid metal to the solid foam, there were formations of pyrolysis products at the reaction site.

A number of physical models were proposed. The physical models advanced that the composition of the mixture of degradation products is a function of the time that the products reside in the interface before escaping out through the coating and sand. Worthy of note is the proposal of Molibog who had advanced that there is a three-phase kinetic zone in the metal-foam interface. Molibog's proposal is shown in figure.

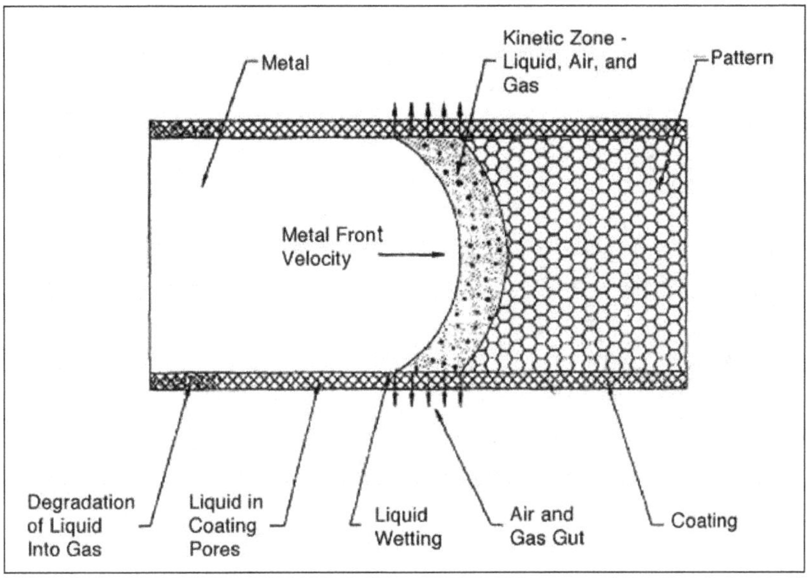

Schematic of the foam degradation process.

From the available literature, it was observed that at the initial stage, as temperature increased, the polymer retained its structure without any changes until the temperature reaches the glass transition temperature (T_g). T_g is the temperature at which the molecules begin to vibrate. This is the initial stage of translational movement.

Immediately the temperature is raised above the T_g the molecules of the polymers increase in vibration and its viscosity reduces. A point of collapse of the polymer beads is reached when the temperature is increased further. Partial depolymerization of the polymer chain is achieved as the temperature proceeds. The products of the depolymerization include monomers, dimers and other oligomers. Usually, with EPS, the beads collapse at a temperature about 120 °C and its volume decreases to 1/40 of its original size. The degradation process known with the EPS occurs by random scission. The C-C bonds contained in the polymer break down at different locations. It has been noted that at 160 °C, the beads which are expanded by team molding to produce the EPS melt, forming viscous substance, consisting of depolymerized products. At various temperatures, the products are formed and they are known to undergo further fragmentations. Finally at about 750 °C, the gaseous products formed contain styrene, toluene (C_7H_8), benzene (C_6H_6), ethylene (C_2H_4), acetylene (C_2H_2) and methane (CH_4). The properties of expanded polystyrene (EPS) and poly methyl methacrylate (PMMA) thermal degradation are summarized in Table.

Table: The characteristics of EPS thermal degradation.

Thermal properties	EPS	PMMA
Glass transition temperature (°C)	80 to 100	105
Collapse temperature (°C)	110 to 120	140 to 200
Melting temperature (°C)	160	260
Starting temperature of volatilization (°C)	275 to 300	250 to 260
Peak volatilization temperature (°C)	400 to 420	370
End volatilization temperature (°C)	460 to 500	420 to 430
Heat of degradation (J/g)	912	842

Rate of vaporization at 750 °C (Kg/s.m.²)	0.77	0.61
Rate of vaporization at 1300 °C (Kg/s.m.²)	0.18	0.31
Gas yield at 750 °C (m³(STP)/Kg)	0.23	0.273
Gas yield at 1300 °C (m³(STP)/Kg)	0.76	0.804
% Viscous residue at 750 °C	61	32
% Viscous residue at 1400 °C	15	3

In iron founding, lustrous carbon defects are known to occur by using EPS as the pattern material. It is to this end that alternative is provided called poly methyl methacrylate (PMMA) as the pattern material. This is capable of eliminating the defects known with EPS. The properties of the PMMA are included in Table. The PMMA is known to undergo degradation by a process called unzipping. Research efforts to eliminate completely or at least reduce the lustrous carbon defects of the EPS in iron founding have yielded positive results by the invention of other polymers such as polyalkylene (PAC) and copolymers of the EPS.

Molding Practices in EPC Process

After the patterns have been produced by injection molding as applied to industrial method or machined out from block of polystyrene, they are assembled with the gating systems design for the patterns and coated with refractory material. Time is given to allow the refractory material to dry. The patterns with the gating systems are then taken and positioned in the molds and supported with green molding sand and rammed carefully. The ramming is necessary to provide rigidity for the molds so that they would not collapse when molten metal is poured. The patterns with the gating systems buried in the molds are not removed like those in traditional sand casting methods. When the liquid metal is poured, the patterns and gating system evaporate leaving only the castings at solidification. The schematic of the pattern-gating system assembly positioned in the mold is shown in Figure. The thermal energy helps foam patterns to decompose and leaving desired dimensional casting products. As the metal replaces the foam pattern, the process involves a series of complex foam reactions: collapse, liquefaction, vaporization, and depolymerization. The degradation products are vented through the coating layer into the surrounding sand.

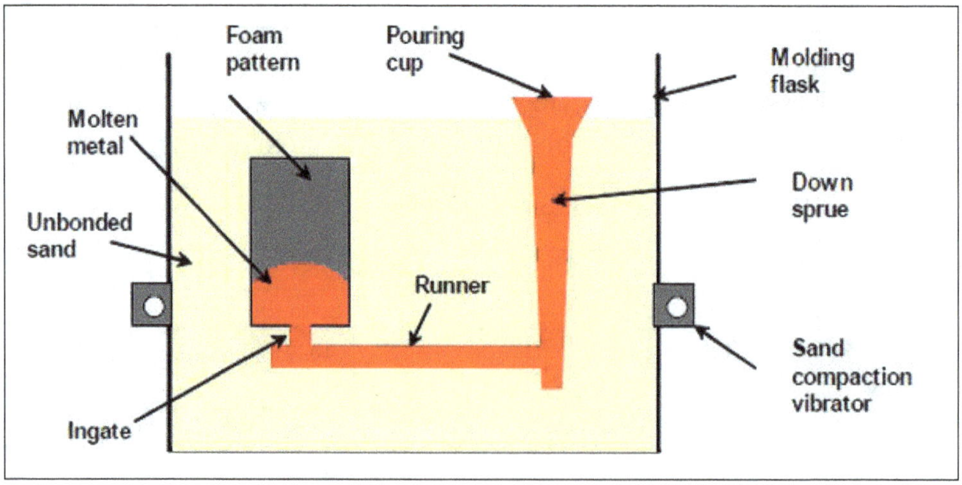

Schematic of molten metal pouring in EPC process.

The quality of castings in the EPC process is strongly affected by the elimination of liquid and gaseous products produced by the foam pattern. If the foam pattern pyrolysis products cannot be effectively eliminated from the casting, they can cause various defects. After the castings cool down, they are shaken out, de-gated, cleaned and inspected for quality. These final procedures are similar to those used in conventional casting process.

Vibration

Vibration of the sand around the pattern to obtain the optimum density and compaction of the mold is a critical feature in the successful production of castings by the EPC. The vibration characteristics are critical since the sand must be caused to flow into internal cavities and undercuts for good compaction. The frequency, amplitude and intensity of vibration, direction of movement impacts, shape of the container and the point at which the vibration is applied all influence the quality of the castings. reported that applying vibration during solidification is an effective way for grain refinement. Vibration can modify the solidified microstructure by promoting nucleation and thus reduce the grain size and lead to a homogenous microstructure. The frequency can be of range 10 to 60 Hz, the amplitude ranges from 0.11 to 0.45 while the intensity which is quantified as force ranges from 80 to 12,000 N.

Characteristic Behaviour of Molten Metal in Prepared Molds

To eliminate mold erosion in sand molding processes, it is important that gating and feeding system is designed and constructed for the particular casting that would be produced. It is the gating system that will eliminate turbulence that may occur as a result of the pouring of the liquid metal into the prepared mold. It enters the mold by gravity.

In EPC, immediately the liquid metal contacts the buried polystyrene foam, it begins to decompose and violent gas escapes from the mold through vents that have been provided. Physical and chemical reactions begin to take place inside the mold. The gas evolving develops a back pressure to resist the advancing liquid metal; by this a kinetic zone is established. The molten metal displaces the foam because of its high temperature; resulting in pyrolysis products like monomers and emissions toxic to human. All the products formed in the mold have been established as potential causes of defects in castings obtained by EPC process. Attempts have been made to study comprehensively the behaviour of the liquid metal poured into EPC molds. Some researchers have employed x-rays to study the physico-chemical reactions taking place at the sites. Attempts have been made to model and simulate the interactions that have been known to occur for possible predictions. However, difficulty still exists to quantitatively/qualitatively represent that behaviour of the fluid in the mold.

Pouring Temperature

This refers to the temperature at which the liquid metal is poured into the mold. It is one of the pouring material variables. By experimental determination, it has been established that pouring temperature has significant effects on the mechanical properties of the castings obtained by sand molding processes, EPC inclusive. Usually, it is at high temperature that the casting material is melted in the furnace. Once the casting material melts, the prepared mold is poured. By experiment, it is determined that some aluminum alloys when poured into EPC mold at 650 °C, quality casting is obtained.

Effects of pouring temperature on the mechanical properties and microstructures of casting can be studied by observing a range of values and levels are taken within the range when the casting material is poured at the levels, opportunity is provided to compare results in terms of visual and mechanical examination of the casting obtained. The visual examination reveals some defects on the surface of the casting if they are existing. Surface roughness can as well be measured. The mechanical examination reveals the mechanical properties in terms of impart, tensile, creep, % elongation, hardness and elasticity. These provide evaluation of the casting to know if it is of good quality. Sometimes the effect of the pouring temperature is examined on the grain refinement.

Each casting material has the temperature range at which it must be poured. Therefore the temperature is measured to determine the pouring range. Usually a k-type chromium-nickel thermocouple is employed to take the measurement. It has a grinded junction. Other forms of thermocouple can be used. The pouring temperature because of the high thermal energy that it has is responsible for the dissociation of H_2O in the molding sand, thereby producing H_2 which escapes from the mold and the decomposition of the evaporative patterns used in EPC process.

The Gating System

Molten metal is poured into already prepared mold through a gating system. The gating system designed for a casting is informed by the weight of the casting. There are a number of parts that make up the gating system. Pouring basin is a reservoir at the top of the gating system that receives the stream of molten metal poured from the ladle. Next to the pouring basin is the sprue which usually tapers down to the sprue base well. The sprue usually in the form of a cone has an exit diameter determined by employing the formula for the area of a cone.

The sprue base well is usually taken to be five the size of the cross-sectional area of the sprue exit. The choke informs what the cross-sectional areas of the runner bar and ingates will be by employing the gating ratios. Pressurized and non-pressurized gating ratios are used; pressurized for the ferrous castings and the non-pressurized for the light alloys such as Al-alloys. The runners and ingates convey the molten metal to the cavity of the mold or the polystyrene pattern buried in the mold in the case of EPC Process. Riser is an important part of the gating system. Its function is to feed the casting during solidification so that no shrinkage cavities are formed.

There are many types of gating system designs; top, bottom, side and step. Jain and Rao posited that side gating system enjoys the advantages of both the top and bottom gating designs. However, gating system should be used based on the shape of component to be produced. For example the bottom design is best for the cylinder. Figures show the top, bottom and side or parting line gating designs. These gating systems must be designed, constructed and assembled with the patterns and positioned in the mold ready for the pouring of the liquid metal. Rao provided an insight into the pouring rate of molten metals with respect to their weights poured into already prepared mold. This assists the foundry engineer to determine the rate of flow of liquid metal which can ultimately be used to determine the pouring time. Table gives the pouring rates of some metals.

Table: Pouring rate of metals.

Metal	Pouring rate in Kg/s for castings of mass, Kg			
	Up to 10	10–50	50–100	100–500
Pig iron	1.1	1.5–2.0	3.0–4.0	3.5–6.0

Steel	1.2–1.4	1.9–2.5	4.0–5.0	4.5–7.0
Aluminum alloys	0.25–0.3	0.5–0.7	1.0–1.3	1.2–2.0

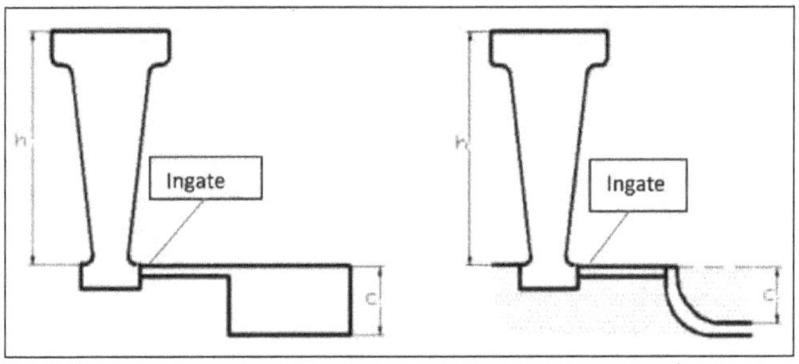

Top gating system.

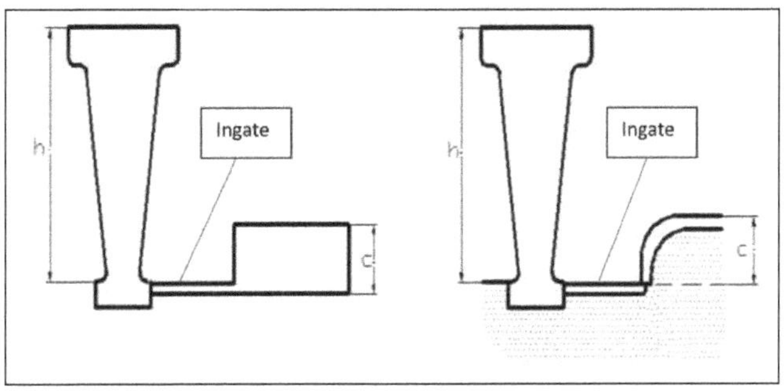

Bottom gating system.

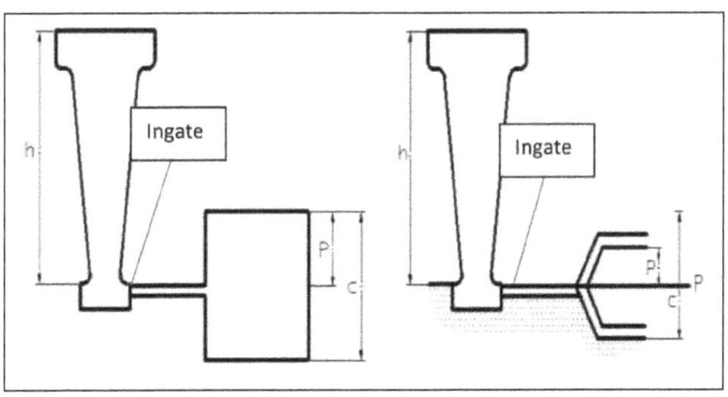

Parting line gating system.

Pouring and Pouring Equipment

To achieving soundness in metal casting, more importantly in EPC Process that is susceptible to many variables, pouring of liquid metal into the sand mold is carefully done. In pouring, the liquid metal must be clean and free from slag. The pouring is done such that air aspiration and turbulence are excluded. By this, mold erosion is avoided and defect like metal penetration does not occur. Sound castings are then obtained. The pouring equipment which is the ladle should not be too heavy for the foundry man to carry, otherwise, a mechanical system should be used to carry out the pouring to prevent accident which may occur as a result of inability of the foundry Introduction.

Lost-foam Casting

Lost-foam casting (LFC) is a type of evaporative-pattern casting process that is similar to investment casting except foam is used for the pattern instead of wax. This process takes advantage of the low boiling point of polymer foams to simplify the investment casting process by removing the need to melt the wax out of the mold.

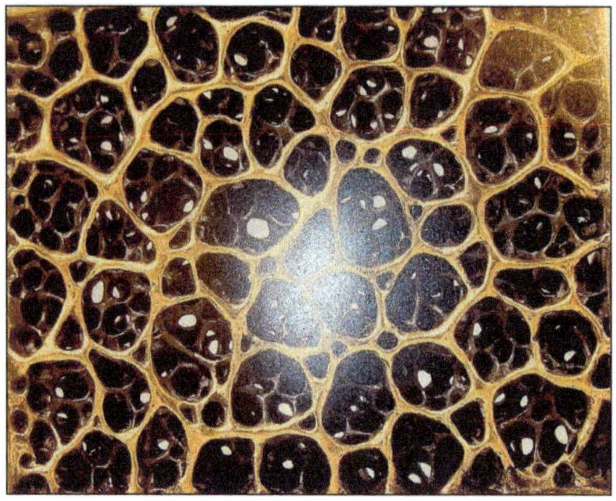

Fragment of artistic bronze casting cell.

Process

First, a pattern is made from polystyrene foam, which can be done by many different ways. For small volume runs the pattern can be hand cut or machined from a solid block of foam; if the geometry is simple enough it can even be cut using a hot-wire foam cutter. If the volume is large, then the pattern can be mass-produced by a process similar to injection molding. Pre-expanded beads of polystyrene are injected into a preheated aluminum mold at low pressure. Steam is then applied to the polystyrene which causes it to expand more to fill the die. The final pattern is approximately 97.5% air and 2.5% polystyrene. Pre-made pouring basins, runners, and risers can be hot glued to the pattern to finish it.

Next, the foam cluster is coated with ceramic investment, also known as the refractory coating, via dipping, brushing, spraying or flow coating. After the coating dries, the cluster is placed into a flask and backed up with un-bonded sand which is compacted using a vibration table. The refractory coating captures all of the detail in the foam model and creates a barrier between the smooth foam surface and the coarse sand surface. Secondly it controls permeability, which allows the gas created by the vaporized foam pattern to escape through the coating and into the sand. Controlling permeability is a crucial step to avoid sand erosion. Finally, it forms a barrier so that molten metal does not penetrate or cause sand erosion during pouring. Once the sand is compacted, the mold is ready to be poured. Automatic pouring is commonly used in LFC, as the pouring process is significantly more critical than in conventional foundry practice.

There is no bake-out phase, as for lost-wax. The melt is poured directly into the foam-filled mold, burning out the foam as it pours. As the foam is of low density, the waste gas produced by this is relatively small and can escape through mold permeability, as for the usual outgassing control.

Commonly cast metals include cast irons, aluminium alloys, steels, and nickel alloys; less frequently stainless steels and copper alloys are also cast. The size range is from 0.5 kg (1.1 lb) to several tonnes (tons). The minimum wall thickness is 2.5 mm (0.098 in) and there is no upper limit. Typical surface finishes are from 2.5 to 25 µm (100 to 1000 µin) RMS. Typical linear tolerances are ±0.005 mm/mm (0.005 in/in).

Advantages and Disadvantages

This casting process is advantageous for very complex castings that would regularly require cores. It is also dimensionally accurate, maintains an excellent surface finish, requires no draft, and has no parting lines so no flash is formed. The un-bonded sand of lost foam casting can be much simpler to maintain than green sand and resin bonded sand systems. Lost foam is generally more economical than investment casting because it involves fewer steps. Risers are not usually required due to the nature of the process; because the molten metal vaporizes the foam the first metal into the mold cools more quickly than the rest, which results in natural directional solidification. Foam is easy to manipulate, carve and glue, due to its unique properties. The flexibility of LFC often allows for consolidating the parts into one integral component; other forming processes would require the production of one or more parts to be assembled.

The two main disadvantages are that pattern costs can be high for low volume applications and the patterns are easily damaged or distorted due to their low strength. If a die is used to create the patterns there is a large initial cost.

Full-mold Casting

In full-mold casting both foam and sand are used to create the mold. The mold is first made out of foam and then cut into the desired or size and shape for the intended application. Next, the mold is coated with heat-resistant sand that binds to its foam surface.

After the mold has been prepared, molten metal is poured into the pattern's cavity. With molten metal reaching 2,500 degrees Fahrenheit or hotter, it's able to vaporize foam on contact. As the foam vaporizes, it creates a new object in the same shape and size as the mold pattern. The area in which the foam was located prior to being vaporized is replaced with metal. Once the molten metal cools and hardens, the newly created object is removed from the mold pattern.

Full-mold casting is most commonly used to produce automotive components like exhaust manifolds, cylinder heads and brake rotors. Of course, full-mold casting can be used for countless other applications. It's often preferred over other molding processes because of the unique advantages it offers.

Advantages of Full-mold Casting

- Supports complex shapes.
- Highly accurate.
- Highly efficient (produces little or no waste).
- Costs less than traditional wax-based casting processes.
- Doesn't require draft or flash.
- Supports a variety of sizes, ranging up to several tons.
- Can be used with molten aluminum, steel, iron, copper, alloys and many other metals.

Disadvantages of Full-mold Casting

- While relatively inexpensive when performed for high-volume applications, such as the mass-production of goods, full-mold casting typically costs more than other casting processes for low-volume applications.
- Mold patterns are susceptible to cracking and other forms of damage.

Full-mold Casting vs Sand Casting and Lost-foam Casting

Full-mold casting isn't the same as sand casting or lost-foam casting. Rather, it uses elements from both of these alternative evaporative-pattern casting processes. With sand casting, the molds are made entirely of sand. With lost-foam casting, the molds are made entirely of foam. Full-mold casting, on the other hand, uses a sand-covered foam mold.

INVESTMENT CASTING

Investment casting is an industrial process based on *lost-wax casting*, one of the oldest known metal-forming techniques. The term "lost-wax casting" can also refer to modern investment casting processes.

Investment casting has been used in various forms for the last 5,000 years. In its earliest forms, beeswax was used to form patterns necessary for the casting process. Today, more advanced waxes, refractory materials and specialist alloys are typically used for making patterns. Investment casting is valued for its ability to produce components with accuracy, repeatability, versatility and integrity in a variety of metals and high-performance alloys.

The fragile wax patterns must withstand forces encountered during the mold making. Much of the wax used in investment casting can be reclaimed and reused. Lost-foam casting is a modern form of investment casting that eliminates certain steps in the process.

Investment casting derives its name from the pattern being invested (surrounded) with a refractory material. Many materials are suitable for investment casting; examples are stainless steel alloys, brass, aluminium, carbon steel and glass. The material is poured into a cavity in a refractory material that is an exact duplicate of the desired part. Due to the hardness of refractory materials used, investment casting can produce products with exceptional surface qualities, which can reduce the need for secondary machine processes.

Water glass and silica sol investment casting are the two primary investment casting methods nowadays. The main differences are the surface roughness and cost of casting. Water glass method dewaxes into the high-temperature water, and the ceramic mold is made of water glass quartz sand. Silica sol method dewaxes into the flash fire, and silica sol zircon sand makes the ceramic mold. Silica sol method costs more but has the better surface than the water glass method.

The process can be used for both small castings of a few ounces and large castings weighing several hundred pounds. It can be more expensive than die casting or sand casting, but per-unit costs decrease with large volumes. Investment casting can produce complicated shapes that would be difficult or impossible with other casting methods. It can also produce products with exceptional surface qualities and low tolerances with minimal surface finishing or machining required.

Inlet-outlet cover of a valve for a nuclear power station produced using investment casting.

Process

A wax pattern used to create a jet engine turbine blade.

Castings can be made from an original wax model (the direct method) or from wax replicas of an original pattern that need not be made from wax (the indirect method). The following steps describe the indirect process, which can take two to seven days to complete:

- Produce a master pattern: An artist or mould-maker creates an original pattern from wax, clay, wood, plastic, or another material. In recent years the production of patterns using 3D printing has become popular using either standard PLA filament or custom made 'casting wax' filament, in which case one goes directly to step 5.
- Create a mould: A mould, known as the *master die*, is made to fit the master pattern. If the master pattern was made from steel, the master die can be cast directly from the pattern using metal with a lower melting point. Rubber moulds can also be cast directly from the master pattern. Alternatively, a master die can be machined independently—without creating a master pattern.
- Produce wax patterns: Although called *wax patterns*, pattern materials may also include plastic and frozen mercury. Wax patterns can be produced in one of two ways. In one process, the wax is poured into the mold and swished around until an even coating, usually about 3 mm (0.12 in) thick, covers the inner surface of the mould. This is repeated until

the desired pattern thickness is reached. Another method involves filling the entire mould with molten wax and letting it cool as a solid object. If a core is required, there are two options: soluble wax or ceramic. Soluble wax cores are designed to melt out of the investment coating with the rest of the wax pattern; ceramic cores are removed after the product has hardened.

- Assemble wax patterns: Multiple wax patterns can be created and assembled into one large pattern to be cast in one batch pour. In this situation, patterns are attached to a wax sprue to create a pattern cluster, or *tree*. To attach patterns, a heating tool is used to slightly melt designated wax surfaces, which are then pressed against each other and left to cool and harden. As many as several hundred patterns can be assembled into a tree. Wax patterns can also be *chased*, which means parting lines or flashings are rubbed out using the heated metal tool. Finally, patterns are *dressed* (by removing imperfections) to look like finished pieces.

- Apply investment materials: The ceramic mould, known as the *investment*, is produced by repeating a series of steps—coating, stuccoing, and hardening—until a desired thickness is achieved. *Coating* involves dipping a pattern cluster into a slurry of fine refractory material and then draining to create a uniform surface coating. Fine materials are used in this first step, also called a *prime coat*, to preserve fine details from the mould. *Stuccoing* applies coarse ceramic particles by dipping patterns into a fluidised bed, placing it in a rainfall-sander, or by applying materials by hand. *Hardening* allows coatings to cure. These steps are repeated until the investment reaches its required thickness—usually 5 to 15 mm (0.2 to 0.6 in). Investment moulds are left to dry completely, which can take 16 to 48 hours. Drying can be accelerated by applying a vacuum or minimizing environmental humidity. Investment moulds can also be created by placing the pattern clusters into a flask and then pouring liquid investment material from above. The flask is then vibrated to allow entrapped air to escape and help the investment material fill any small voids. Common refractory materials used to create the investments are: silica, zircon, various aluminium silicates, and alumina. Silica is usually used in the fused silica form, but sometimes quartz is used because it is less expensive. Aluminium silicates are a mixture of alumina and silica, where commonly used mixtures have an alumina content from 42 to 72%; at 72% alumina the compound is known as mullite. During the primary coat(s), zircon-based refractories are commonly used, because zirconium is less likely to react with the molten metal. Prior to silica, a mixture of plaster and ground up old molds (chamotte) was used. The binders used to hold the refractory material in place include: ethyl silicate (alcohol-based and chemically set), colloidal silica (water-based, also known as silica sol, set by drying), sodium silicate, and a hybrid of these controlled for pH and viscosity.

- Dewax: Once ceramic moulds have fully cured, they are turned upside-down and placed in a furnace or autoclave to melt out and/or vaporize the wax. Most shell failures occur at this point because the waxes used have a thermal expansion coefficient that is much greater than the investment material surrounding it—as the wax is heated it expands and introduces stress. To minimize these stresses the wax is heated as rapidly as possible so that outer wax surfaces can melt and drain quickly, making space for the rest of the wax to expand. In certain situations, holes may be drilled into the mold before heating to help reduce these stresses. Any wax that runs out of the mold is usually recovered and reused.

- Burnout preheating: The mold is then subjected to a *burnout*, which heats the mold to between 870 °C and 1095 °C to remove any moisture and residual wax, and to sinter the mold. Sometimes this heating is also used to preheat the mold before pouring, but other times the mould is allowed to cool so that it can be tested. Preheating allows the metal to stay liquid longer so that it can better fill all mould details and increase dimensional accuracy. If the mold is left to cool, any cracks found can be repaired with ceramic slurry or special cements.
- Pouring: The investment mold is then placed open-side up into a tub filled with sand. The metal may be gravity poured or forced by applying positive air pressure or other forces. Vacuum casting, tilt casting, pressure assisted pouring and centrifugal casting are methods that use additional forces and are especially useful when moulds contain thin sections that would be otherwise be difficult to fill.
- Divesting: The shell is hammered, media blasted, vibrated, waterjeted, or chemically dissolved (sometimes with liquid nitrogen) to release the casting. The sprue is cut off and recycled. The casting may then be cleaned up to remove signs of the casting process, usually by grinding.
- Finishing: After grinding, the completed casting is then subject to finishing. This usually goes further than grinding, with impurities and negatives being removed via hand tooling and welding. In the case that the part needs additional straightening, this process is usually carried out by hydraulic straightening presses, which bring the product in line with its tolerances.

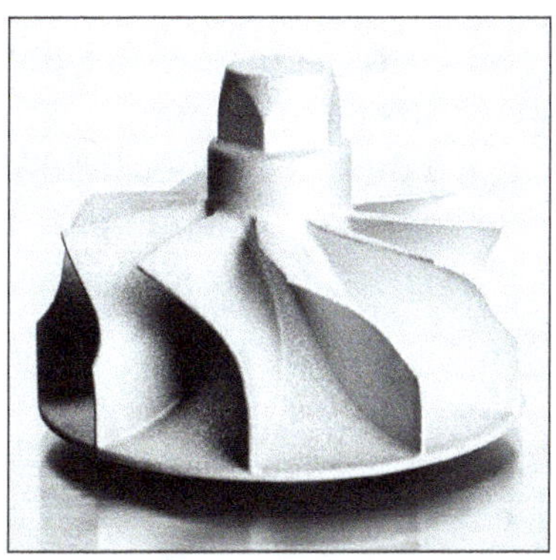

The investment shell for casting a turbocharger rotor.

A view of the interior investment shows the smooth surface finish and high level of detail.

The completed workpiece.

Advantages

- Excellent surface finish.
- High dimensional accuracy.
- Extremely intricate parts are castable.
- Almost any metal can be cast.
- No flash or parting lines.

Disadvantages

The main disadvantage is the overall cost, especially for short-run productions. Some of the reasons for the high cost include specialized equipment, costly refractories, and binders, many operations to make a mould, a lot of labor is needed and occasional minute defects occur. However, the cost is still less than producing the same part by machining from bar stock; for example, gun manufacturing has moved to investment casting to lower costs of producing pistols.

Additionally:

- It can be difficult to cast objects requiring cores.
- This process is expensive, is usually limited to small casting, and presents some difficulties where cores are involved.
- Holes cannot be smaller than 1/16 in. (1.6 mm) and should be no deeper than about 1.5 times the diameter.
- Investment castings require longer production cycles compared to other casting processes.

Counter-gravity Casting

The variation on the gravity pouring technique is to fill the mold using a vacuum. A common form of this is called the *Hitchiner* process after the Hitchiner Manufacturing Company that invented the technique. In this technique, the mold has a downward fill pipe that is lowered into the melt. A vacuum draws the melt into the cavity; when the important parts have solidified, the vacuum is released, and the unused material leaves the mold. The technique can use substantially less material than gravity pouring because the sprue and some gating need not solidify.

This technique is more metal efficient than traditional pouring because less material solidifies in the gating system. Gravity pouring only has a 15 to 50% metal yield compared to 60 to 95% for counter-gravity pouring. There is also less turbulence, so the gating system can be simplified since it does not have to control turbulence. The metal is drawn from below the top of the pool, so the metal is free from dross and slag (which are lower density (lighter) and float to the top of the pool). The pressure differential helps the metal flow into every intricacy of the mold. Finally, lower temperatures can be used, which improves the grain structure.

This process is also used to cast refractory ceramics under the term vacuum casting.

Vacuum Pressure Casting

Vacuum pressure casting (VPC), properly referred to as *vacuum assist direct pour*, uses gas pressure and a vacuum to improve the quality of the casting and minimize porosity. Typically VPC machines consist of an upper and a lower chamber—the upper chamber, or melting chamber, housing the crucible, and the lower casting chamber housing the investment mould. Both chambers are connected via a small hole containing a stopper. A vacuum is pulled in the lower chamber, while pressure is applied in the upper, and then the stopper is removed. This creates the greatest pressure differential to fill the molds. The most common materials for *vacuum casting process* are the high nickel-based alloy and super alloys. Turbocharger products are a common applications for this casting process, though it is also regularly used in the manufacture of silver and gold jewellery.

Investment casting is used with almost any castable metal. However, aluminium alloys, copper alloys, and steel are the most common. In industrial use, the size limits are 3 g (0.1 oz) to several hundred kilograms. The cross-sectional limits are 0.6 mm (0.024 in) to 75 mm (3.0 in). Typical tolerances are 0.1 mm for the first 25 mm (0.005 in for the first inch) and 0.02 mm for the each additional centimeter (0.002 in for each additional inch). A standard surface finish is 1.3–4 micrometres (50–125 μ) RMS.

Applications

Investment casting is used in the aerospace and power generation industries to produce turbine blades with complex shapes or cooling systems. Blades produced by investment casting can include single-crystal (SX), directionally solidified (DS), or conventional equiaxed blades.

Investment casting is also widely used by firearms manufacturers to fabricate firearm receivers, triggers, hammers, and other precision parts at low cost.

Karsten Solheim famously revolutionized golf club design through his company PING by incorporating investment casting for the first time for clubheads. Quickly the process became an industry standard to allow weight distribution around the perimeter of the clubhead.

Other industries that use standard investment-cast parts include military, medical, commercial and automotive.

Unveiling the titanium integral space bus satellite. The sacrificial mold for the investment casting was 3D-printed with integral cable routing and toroidal propellant tank.

With the increased availability of higher-resolution 3D printers, 3D printing has begun to be used to make much larger sacrificial molds used in investment casting. Planetary Resources has used the technique to print the mold for a new small satellite, which is then dipped in ceramic to form the investment cast for a titanium space bus with integral propellant tank and embedded cable routing.

SAND CASTING

Sand casting, also known as sand molded casting, is a metal casting process characterized by using sand as the mold material. The term "sand casting" can also refer to an object produced via the sand casting process. Sand castings are produced in specialized factories called foundries. Over 60% of all metal castings are produced via sand casting process.

Cope & drag (top and bottom halves of a sand mold), with cores in place on the drag.

Molds made of sand are relatively cheap, and sufficiently refractory even for steel foundry use. In addition to the sand, a suitable bonding agent (usually clay) is mixed or occurs with the sand. The mixture is moistened, typically with water, but sometimes with other substances, to develop the strength and plasticity of the clay and to make the aggregate suitable for molding. The sand is typically contained in a system of frames or mold boxes known as a flask. The mold cavities and gate system are created by compacting the sand around models called patterns, by carving directly into the sand, or by 3D printing.

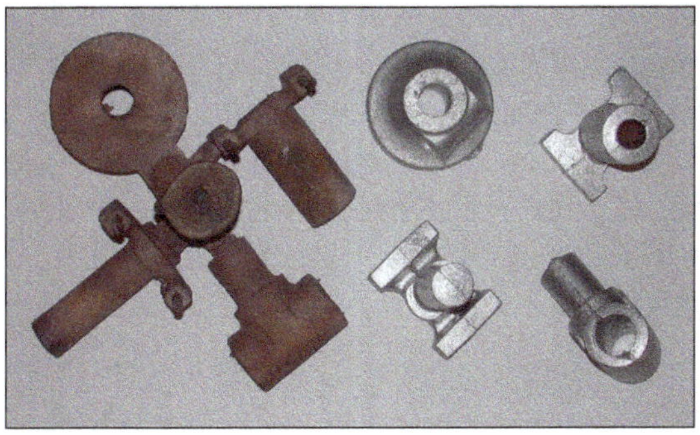

Two sets of castings (bronze and aluminium) from sand mold.

Basic Process

There are six steps in this process:

- Place a pattern in sand to create a mold.
- Incorporate the pattern and sand in a gating system.
- Remove the pattern.
- Fill the mold cavity with molten metal.
- Allow the metal to cool.
- Break away the sand mold and remove the casting.

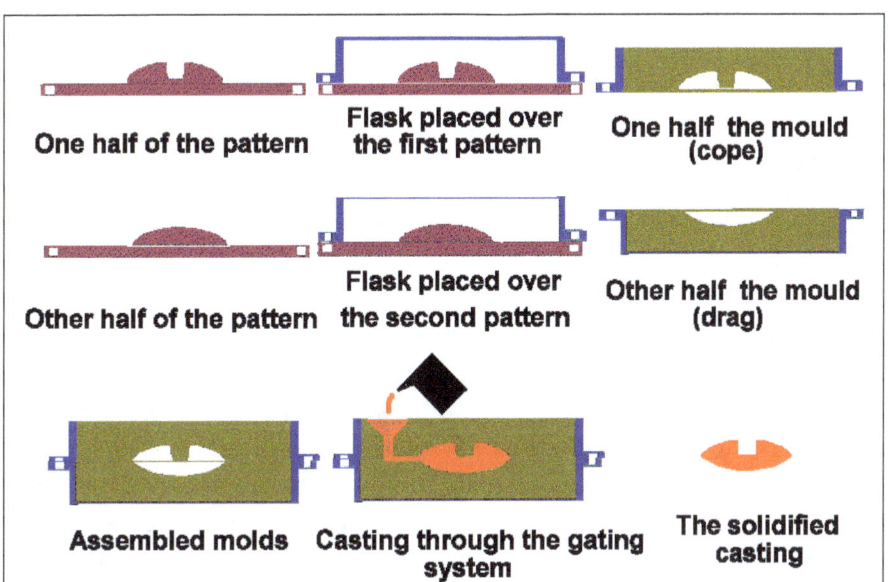

Components

Patterns

From the design, provided by a designer, a skilled *pattern maker* builds a *pattern* of the object to be produced, using wood, metal, or a plastic such as expanded polystyrene. Sand can be ground, swept or strickled into shape. The metal to be cast will contract during solidification, and this may be non-uniform due to uneven cooling. Therefore, the pattern must be slightly larger than the finished product, a difference known as *contraction allowance*. Different scaled rules are used for different metals, because each metal and alloy contracts by an amount distinct from all others. Patterns also have core prints that create registers within the molds into which are placed sand cores. Such cores, sometimes reinforced by wires, are used to create under-cut profiles and cavities which cannot be molded with the cope and drag, such as the interior passages of valves or cooling passages in engine blocks.

Paths for the entrance of metal into the mold cavity constitute the runner system and include the sprue, various feeders which maintain a good metal 'feed', and in-gates which attach the runner system to the casting cavity. Gas and steam generated during casting exit through the permeable sand or via risers, which are added either in the pattern itself, or as separate pieces.

Tools

In addition to patterns, the sand molder could also use tools to create the holes.

Sand Molding tools and books.

Molding Box and Materials

A multi-part molding box (known as a casting flask, the top and bottom halves of which are known respectively as the cope and drag) is prepared to receive the pattern. Molding boxes are made in segments that may be latched to each other and to end closures. For a simple object—flat on one side—the lower portion of the box, closed at the bottom, will be filled with a molding sand. The sand is packed in through a vibratory process called ramming, and in this case, periodically screeded level. The surface of the sand may then be stabilized with a sizing compound. The pattern is placed on the sand and another molding box segment is added. Additional sand is rammed over and around the pattern. Finally a cover is placed on the box and it is turned and unlatched, so that the halves of the mold may be parted and the pattern with its sprue and vent patterns removed. Additional sizing may be added and any defects introduced by the removal of the pattern are corrected. The box is closed again. This forms a "green" mold which must be dried to receive the hot metal. If the mold is not sufficiently dried a steam explosion can occur that can throw molten metal about. In some cases, the sand may be oiled instead of moistened, which makes casting possible without waiting for the sand to dry. Sand may also be bonded by chemical binders, such as furane resins or amine-hardened resins.

Additive manufacturing can be used in the sand mold preparation, so that instead of the sand mold being formed via packing sand around a pattern, it is 3D-printed. This can reduce lead times for casting by obviating patternmaking. Besides replacing older methods, additive can also complement them in hybrid models, such as making a variety of AM-printed cores for a cavity derived from a traditional pattern.

Chills

To control the solidification structure of the metal, it is possible to place metal plates, *chills*, in the mold. The associated rapid local cooling will form a finer-grained structure and may form a somewhat harder metal at these locations. In ferrous castings, the effect is similar to quenching metals in forge work. The inner diameter of an engine cylinder is made hard by a chilling core. In other metals, chills may be used to promote directional solidification of the casting. In controlling the way a casting freezes, it is possible to prevent internal voids or porosity inside castings.

Cores

To produce cavities within the casting—such as for liquid cooling in engine blocks and cylinder heads—negative forms are used to produce *cores*. Usually sand-molded, cores are inserted into the casting box after removal of the pattern. Whenever possible, designs are made that avoid the use of cores, due to the additional set-up time, mass and thus greater cost.

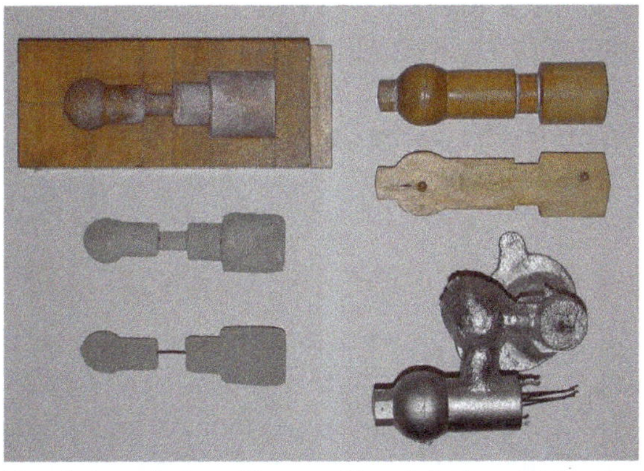

Left: Corebox, with resulting (wire reinforced) cores directly below. Right: Pattern (used with the core) and the resulting casting below (the wires are from the remains of the core).

With a completed mold at the appropriate moisture content, the box containing the sand mold is then positioned for filling with molten metal—typically iron, steel, bronze, brass, aluminium, magnesium alloys, or various pot metal alloys, which often include lead, tin, and zinc. After being filled with liquid metal the box is set aside until the metal is sufficiently cool to be strong. The sand is then removed, revealing a rough casting that, in the case of iron or steel, may still be glowing red. In the case of metals that are significantly heavier than the casting sand, such as iron or lead, the casting flask is often covered with a heavy plate to prevent a problem known as *floating the mold.*

Floating the mold occurs when the pressure of the metal pushes the sand above the mold cavity out of shape, causing the casting to fail.

After casting, the cores are broken up by rods or shot and removed from the casting. The metal from the sprue and risers is cut from the rough casting. Various heat treatments may be applied to relieve stresses from the initial cooling and to add hardness—in the case of steel or iron, by quenching in water or oil. The casting may be further strengthened by surface compression treatment—like shot peening—that adds resistance to tensile cracking and smooths the rough surface. And when high precision is required, various machining operations (such as milling or boring) are made to finish critical areas of the casting. Examples of this would include the boring of cylinders and milling of the deck on a cast engine block.

Design Requirements

The part to be made and its pattern must be designed to accommodate each stage of the process, as it must be possible to remove the pattern without disturbing the molding sand and to have proper locations to receive and position the cores. A slight taper, known as draft, must be used on surfaces perpendicular to the parting line, in order to be able to remove the pattern from the mold. This requirement also applies to cores, as they must be removed from the core box in which they are formed. The sprue and risers must be arranged to allow a proper flow of metal and gasses within the mold in order to avoid an incomplete casting. Should a piece of core or mold become dislodged it may be embedded in the final casting, forming a *sand pit*, which may render the casting unusable. Gas pockets can cause internal voids. These may be immediately visible or may only be revealed after extensive machining has been performed. For critical applications, or where the cost of wasted effort is a factor, non-destructive testing methods may be applied before further work is performed.

Processes

In general, we can distinguish between two methods of sand casting; the first one using *green sand* and the second being the *air set* method.

Green Sand

These castings are made using sand molds formed from "wet" sand which contains water and organic bonding compounds, typically referred to as clay. The name "Green Sand" comes from the fact that the sand mold is not "set", it is still in the "green" or uncured state even when the metal is poured in the mould. Green sand is not green in color, but "green" in the sense that it is used in a wet state (akin to green wood). Contrary to what the name suggests, "green sand" is not a type of sand on its own (that is, not greensand in the geologic sense), but is rather a mixture of:

- Silica sand (SiO_2), chromite sand ($FeCr_2O_4$), or zircon sand ($ZrSiO_4$), 75 to 85%, sometimes with a proportion of olivine, staurolite, or graphite.
- Bentonite (clay), 5 to 11%.
- Water, 2 to 4%.

- Inert sludge 3 to 5%.
- Anthracite (0 to 1%).

There are many recipes for the proportion of clay, but they all strike different balances between moldability, surface finish, and ability of the hot molten metal to degas. Coal, typically referred to in foundries as sea-coal, which is present at a ratio of less than 5%, partially combusts in the presence of the molten metal, leading to offgassing of organic vapors. Green sand casting for non-ferrous metals does not use coal additives, since the CO created does not prevent oxidation. Green sand for aluminum typically uses olivine sand (a mixture of the minerals forsterite and fayalite, which is made by crushing dunite rock).

The choice of sand has a lot to do with the temperature at which the metal is poured. At the temperatures that copper and iron are poured, the clay gets inactivated by the heat, in that the montmorillonite is converted to illite, which is a non-expanding clay. Most foundries do not have the very expensive equipment to remove the burned out clay and substitute new clay, so instead, those that pour iron typically work with silica sand that is inexpensive compared to the other sands. As the clay is burned out, newly mixed sand is added and some of the old sand is discarded or recycled into other uses. Silica is the least desirable of the sands, since metamorphic grains of silica sand have a tendency to explode to form sub-micron sized particles when thermally shocked during pouring of the molds. These particles enter the air of the work area and can lead to silicosis in the workers. Iron foundries spend a considerable effort on aggressive dust collection to capture this fine silica. The sand also has the dimensional instability associated with the conversion of quartz from alpha quartz to beta quartz at 680 °C (1250 °F). Often, combustible additives such as wood flour are added to create spaces for the grains to expand without deforming the mold. Olivine, chromite, etc. are therefore used because they do not have a phase transition that causes rapid expansion of the grains, as well as offering greater density, which cools the metal faster, producing finer grain structures in the metal. Since they are not metamorphic minerals, they do not have the polycrystals found in silica, and subsequently do not form hazardous sub-micron sized particles.

Air Set Method

The *air set* method uses dry sand bonded with materials other than clay, using a fast curing adhesive. The latter may also be referred to as no bake mold casting. When these are used, they are collectively called "air set" sand castings to distinguish them from "green sand" castings. Two types of molding sand are natural bonded (bank sand) and synthetic (lake sand); the latter is generally preferred due to its more consistent composition.

With both methods, the sand mixture is packed around a *pattern*, forming a mold cavity. If necessary, a temporary plug is placed in the sand and touching the pattern in order to later form a channel into which the casting fluid can be poured. Air-set molds are often formed with the help of a casting flask having a top and bottom part, termed the cope and drag. The sand mixture is tamped down as it is added around the pattern, and the final mold assembly is sometimes vibrated to compact the sand and fill any unwanted voids in the mold. Then the pattern is removed along with the channel plug, leaving the mold cavity. The casting liquid (typically molten metal) is then poured into the mold cavity. After the metal has solidified and cooled, the casting is separated from the sand mold. There is typically no mold release agent, and the mold is generally destroyed in the removal process.

The accuracy of the casting is limited by the type of sand and the molding process. Sand castings made from coarse green sand impart a rough texture to the surface, and this makes them easy to identify. Castings made from fine green sand can shine as cast but are limited by the depth to width ratio of pockets in the pattern. Air-set molds can produce castings with smoother surfaces than coarse green sand but this method is primarily chosen when deep narrow pockets in the pattern are necessary, due to the expense of the plastic used in the process. Air-set castings can typically be easily identified by the burnt color on the surface. The castings are typically shot blasted to remove that burnt color. Surfaces can also be later ground and polished, for example when making a large bell. After molding, the casting is covered with a residue of oxides, silicates and other compounds. This residue can be removed by various means, such as grinding, or shot blasting.

During casting, some of the components of the sand mixture are lost in the thermal casting process. Green sand can be reused after adjusting its composition to replenish the lost moisture and additives. The pattern itself can be reused indefinitely to produce new sand molds. The sand molding process has been used for many centuries to produce castings manually. Since 1950, partially automated casting processes have been developed for production lines.

Air Set Molding has many advantages. The process is designed to meet the growing demand of casting design engineers for the foundry industry and is best suited for larger, heavier and more complex castings. This process provides an excellent as-cast surface finish for products that require high aesthetic standards.

Cold Box

Uses organic and inorganic binders that strengthen the mold by chemically adhering to the sand. This type of mold gets its name from not being baked in an oven like other sand mold types. This type of mold is more accurate dimensionally than green-sand molds but is more expensive. Thus it is used only in applications that necessitate it.

No-bake Molds

No-bake molds are expendable sand molds, similar to typical sand molds, except they also contain a quick-setting liquid resin and catalyst. Rather than being rammed, the molding sand is poured into the flask and held until the resin solidifies, which occurs at room temperature. This type of molding also produces a better surface finish than other types of sand molds. Because no heat is involved it is called a cold-setting process. Common flask materials that are used are wood, metal, and plastic. Common metals cast into no-bake molds are brass, iron (ferrous), and aluminum alloys.

Vacuum Molding

Vacuum molding (*V-process*) is a variation of the sand casting process for most ferrous and non-ferrous metals, in which unbonded sand is held in the flask with a vacuum. The pattern is specially vented so that a vacuum can be pulled through it. A heat-softened thin sheet (0.003 to 0.008 in (0.076 to 0.203 mm)) of plastic film is draped over the pattern and a vacuum is drawn

(200 to 400 mmHg (27 to 53 kPa)). A special vacuum forming flask is placed over the plastic pattern and is filled with a free-flowing sand. The sand is vibrated to compact the sand and a sprue and pouring cup are formed in the cope. Another sheet of plastic is placed over the top of the sand in the flask and a vacuum is drawn through the special flask; this hardens and strengthens the unbonded sand. The vacuum is then released on the pattern and the cope is removed. The drag is made in the same way (without the sprue and pouring cup). Any cores are set in place and the mold is closed. The molten metal is poured while the cope and drag are still under a vacuum, because the plastic vaporizes but the vacuum keeps the shape of the sand while the metal solidifies. When the metal has solidified, the vacuum is turned off and the sand runs out freely, releasing the casting.

The V-process is known for not requiring a draft because the plastic film has a certain degree of lubricity and it expands slightly when the vacuum is drawn in the flask. The process has high dimensional accuracy, with a tolerance of ±0.010 in for the first inch and ±0.002 in/in thereafter. Cross-sections as small as 0.090 in (2.3 mm) are possible. The surface finish is very good, usually between 150 and 125 rms. Other advantages include no moisture related defects, no cost for binders, excellent sand permeability, and no toxic fumes from burning the binders. Finally, the pattern does not wear out because the sand does not touch it. The main disadvantage is that the process is slower than traditional sand casting so it is only suitable for low to medium production volumes; approximately 10 to 15,000 pieces a year. However, this makes it perfect for prototype work, because the pattern can be easily modified as it is made from plastic.

Fast Mold Making Processes

With the fast development of the car and machine building industry the casting consuming areas called for steady higher productivity. The basic process stages of the mechanical molding and casting process are similar to those described under the manual sand casting process. The technical and mental development however was so rapid and profound that the character of the sand casting process changed radically.

Mechanized Sand Molding

The first mechanized molding lines consisted of sand slingers and/or jolt-squeeze devices that compacted the sand in the flasks. Subsequent mold handling was mechanical using cranes, hoists and straps. After core setting the copes and drags were coupled using guide pins and clamped for closer accuracy. The molds were manually pushed off on a roller conveyor for casting and cooling.

Automatic High Pressure Sand Molding Lines

Increasing quality requirements made it necessary to increase the mold stability by applying steadily higher squeeze pressure and modern compaction methods for the sand in the flasks. In early fifties the high pressure molding was developed and applied in mechanical and later automatic flask lines. The first lines were using jolting and vibrations to pre-compact the sand in the flasks and compressed air powered pistons to compact the molds.

Horizontal Sand Flask Molding

In the first automatic horizontal flask lines the sand was shot or slung down on the pattern in a flask and squeezed with hydraulic pressure of up to 140 bars. The subsequent mold handling including turn-over, assembling, pushing-out on a conveyor were accomplished either manually or automatically. In the late fifties hydraulically powered pistons or multi-piston systems were used for the sand compaction in the flasks. This method produced much more stable and accurate molds than it was possible manually or pneumatically. In the late sixties mold compaction by fast air pressure or gas pressure drop over the pre-compacted sand mold was developed (sand-impulse and gas-impact). The general working principle for most of the horizontal flask line systems is shown on the figure below.

Today there are many manufacturers of the automatic horizontal flask molding lines. The major disadvantages of these systems is high spare parts consumption due to multitude of movable parts, need of storing, transporting and maintaining the flasks and productivity limited to approximately 90–120 molds per hour.

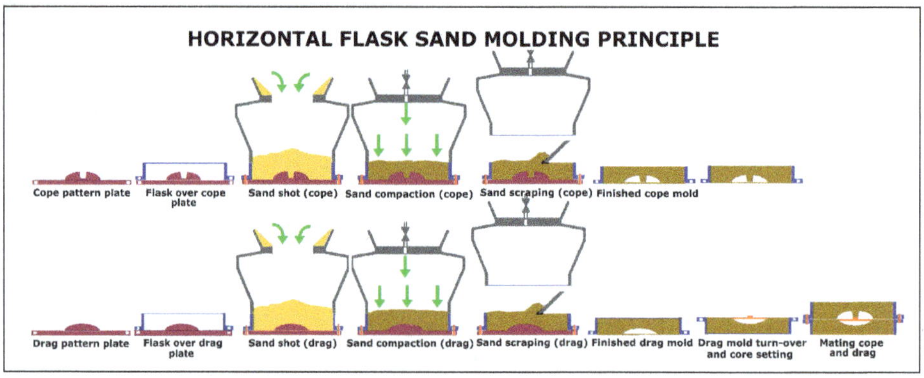

Vertical Sand Flaskless Molding

In 1962, Dansk Industri Syndikat A/S (DISA-DISAMATIC) invented a flask-less molding process by using vertically parted and poured molds. The first line could produce up to 240 complete sand molds per hour. Today molding lines can achieve a molding rate of 550 sand molds per hour and requires only one monitoring operator. Maximum mismatch of two mold halves is 0.1 mm (0.0039 in). Although very fast, vertically parted molds are not typically used by jobbing foundries due to the specialized tooling needed to run on these machines. Cores need to be set with a core mask as opposed to by hand and must hang in the mold as opposed to being set on parting surface.

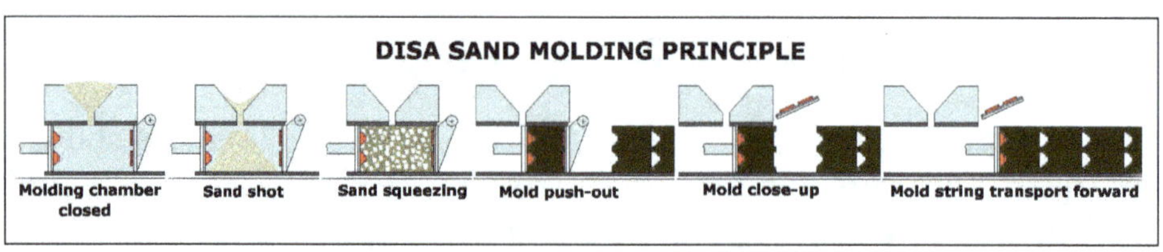

Matchplate Sand Molding

The principle of the matchplate, meaning pattern plates with two patterns on each side of the same plate, was developed and patented in 1910, fostering the perspectives for future sand molding improvements. However, first in the early sixties the American company Hunter Automated

Machinery Corporation launched its first automatic flaskless, horizontal molding line applying the matchplate technology.

The method alike to the DISA's (DISAMATIC) vertical molding is flaskless, however horizontal. The matchplate molding technology is today used widely. Its great advantage is inexpensive pattern tooling, easiness of changing the molding tooling, thus suitability for manufacturing castings in short series so typical for the jobbing foundries. Modern matchplate molding machine is capable of high molding quality, less casting shift due to machine-mold mismatch (in some cases less than 0.15 mm (0.0059 in)), consistently stable molds for less grinding and improved parting line definition. In addition, the machines are enclosed for a cleaner, quieter working environment with reduced operator exposure to safety risks or service-related problems.

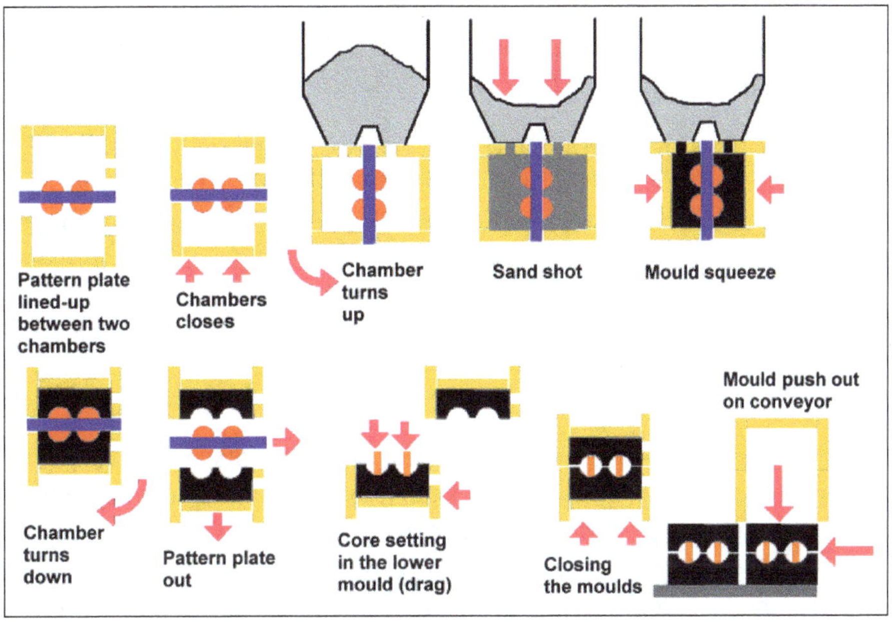

Mold Materials

There are four main components for making a sand casting mold: base sand, a binder, additives, and a parting compound.

Molding Sands

Molding sands, also known as foundry sands, are defined by eight characteristics: refractoriness, chemical inertness, permeability, surface finish, cohesiveness, flowability, collapsibility, and availability/cost:

1. Refractoriness — This refers to the sand's ability to withstand the temperature of the liquid metal being cast without breaking down. For example, some sands only need to withstand 650 °C (1,202 °F) if casting aluminum alloys, whereas steel needs a sand that will withstand 1,500 °C (2,730 °F). Sand with too low refractoriness will melt and fuse to the casting.

2. Chemical inertness — The sand must not react with the metal being cast. This is especially important with highly reactive metals, such as magnesium and titanium.

3. Permeability — This refers to the sand's ability to exhaust gases. This is important because during the pouring process many gases are produced, such as hydrogen, nitrogen, carbon dioxide, and steam, which must leave the mold otherwise casting defects, such as blow holes and gas holes, occur in the casting. Note that for each cubic centimeter (cc) of water added to the mold 16,000 cc of steam is produced.

4. Surface finish — The size and shape of the sand particles defines the best surface finish achievable, with finer particles producing a better finish. However, as the particles become finer (and surface finish improves) the permeability becomes worse.

5. Cohesiveness (or bond) — This is the ability of the sand to retain a given shape after the pattern is removed.

6. Flowability – The ability for the sand to flow into intricate details and tight corners without special processes or equipment.

7. Collapsibility — This is the ability of the sand to be easily stripped off the casting after it has solidified. Sands with poor collapsibility will adhere strongly to the casting. When casting metals that contract a lot during cooling or with long freezing temperature ranges a sand with poor collapsibility will cause cracking and hot tears in the casting. Special additives can be used to improve collapsibility.

8. Availability/cost — The availability and cost of the sand is very important because for every ton of metal poured, three to six tons of sand is required. Although sand can be screened and reused, the particles eventually become too fine and require periodic replacement with fresh sand.

In large castings it is economical to use two different sands, because the majority of the sand will not be in contact with the casting, so it does not need any special properties. The sand that is in contact with the casting is called *facing sand*, and is designed for the casting on hand. This sand will be built up around the pattern to a thickness of 30 to 100 mm (1.2 to 3.9 in). The sand that fills in around the facing sand is called *backing sand*. This sand is simply silica sand with only a small amount of binder and no special additives.

Types of Base Sands

Base sand is the type used to make the mold or core without any binder. Because it does not have a binder it will not bond together and is not usable in this state.

Silica Sand

Silica (SiO_2) sand is the sand found on a beach and is also the most commonly used sand. It is made by either crushing sandstone or taken from natural occurring locations, such as beaches and river beds. The fusion point of pure silica is 1,760 °C (3,200 °F), however the sands used have a lower melting point due to impurities. For high melting point casting, such as steels, a minimum of 98% pure silica sand must be used; however for lower melting point metals, such as cast iron and non-ferrous metals, a lower purity sand can be used (between 94 and 98% pure).

Silica sand is the most commonly used sand because of its great abundance, and, thus, low cost (therein being its greatest advantage). Its disadvantages are high thermal expansion, which can cause casting defects with high melting point metals, and low thermal conductivity, which can lead to unsound casting. It also cannot be used with certain basic metals because it will chemically interact with the metal, forming surface defects. Finally, it releases silica particulates during the pour, risking silicosis in foundry workers.

Olivine Sand

Olivine is a mixture of orthosilicates of iron and magnesium from the mineral dunite. Its main advantage is that it is free from silica, therefore it can be used with basic metals, such as manganese steels. Other advantages include a low thermal expansion, high thermal conductivity, and high fusion point. Finally, it is safer to use than silica, therefore it is popular in Europe.

Chromite Sand

Chromite sand is a solid solution of spinels. Its advantages are a low percentage of silica, a very high fusion point (1,850 °C (3,360 °F)), and a very high thermal conductivity. Its disadvantage is its costliness, therefore it's only used with expensive alloy steel casting and to make cores.

Zircon Sand

Zircon sand is a compound of approximately two-thirds zircon oxide (Zr_2O) and one-third silica. It has the highest fusion point of all the base sands at 2,600 °C (4,710 °F), a very low thermal expansion, and a high thermal conductivity. Because of these good properties it is commonly used when casting alloy steels and other expensive alloys. It is also used as a mold wash (a coating applied to the molding cavity) to improve surface finish. However, it is expensive and not readily available.

Chamotte Sand

Chamotte is made by calcining fire clay (Al_2O_3-SiO_2) above 1,100 °C (2,010 °F). Its fusion point is 1,750 °C (3,180 °F) and has low thermal expansion. It is the second cheapest sand, however it is still twice as expensive as silica. Its disadvantages are very coarse grains, which result in a poor surface finish, and it is limited to dry sand molding. Mold washes are used to overcome the surface finish problems. This sand is usually used when casting large steel workpieces.

Other Materials

Modern casting production methods can manufacture thin and accurate molds—of a material superficially resembling papier-mâché, such as is used in egg cartons, but that is refractory in nature—that are then supported by some means, such as dry sand surrounded by a box, during the casting process. Due to the higher accuracy it is possible to make thinner and hence lighter castings, because extra metal need not be present to allow for variations in the molds. These thin-mold casting methods have been used since the 1960s in the manufacture of cast-iron engine blocks and cylinder heads for automotive applications.

Binders

Binders are added to a base sand to bond the sand particles together (i.e. it is the glue that holds the mold together).

Clay and Water

A mixture of clay and water is the most commonly used binder. There are two types of clay commonly used: bentonite and kaolinite, with the former being the most common.

Oil

Oils, such as linseed oil, other vegetable oils and marine oils, used to be used as a binder, however due to their increasing cost, they have been mostly phased out. The oil also required careful baking at 100 to 200 °C (212 to 392 °F) to cure (if overheated, the oil becomes brittle, wasting the mold).

Resin

Resin binders are natural or synthetic high melting point gums. The two common types used are urea formaldehyde (UF) and phenol formaldehyde (PF) resins. PF resins have a higher heat resistance than UF resins and cost less. There are also cold-set resins, which use a catalyst instead of a heat to cure the binder. Resin binders are quite popular because different properties can be achieved by mixing with various additives. Other advantages include good collapsibility, low gassing, and they leave a good surface finish on the casting.

MDI (methylene diphenyl diisocyanate) is also a commonly used binder resin in the foundry core process.

Sodium Silicate

Sodium silicate [Na_2SiO_3 or $(Na_2O)(SiO_2)$] is a high strength binder used with silica molding sand. To cure the binder, carbon dioxide gas is used, which creates the following reaction:

$$Na_2O(SiO_2) + CO_2 \rightleftharpoons Na_2CO_3 + 2SiO_2 + \text{Heat}$$

The advantage to this binder is that it can be used at room temperature and is fast. The disadvantage is that its high strength leads to shakeout difficulties and possibly hot tears in the casting.

Additives

Additives are added to the molding components to improve: surface finish, dry strength, refractoriness, and "cushioning properties".

Up to 5% of *reducing agents*, such as coal powder, pitch, creosote, and fuel oil, may be added to the molding material to prevent wetting (prevention of liquid metal sticking to sand particles, thus leaving them on the casting surface), improve surface finish, decrease metal penetration, and burn-on defects. These additives achieve this by creating gases at the surface of the mold cavity, which prevent the liquid metal from adhering to the sand. Reducing agents are not used with steel casting, because they can carburize the metal during casting.

Up to 3% of "cushioning material", such as wood flour, saw dust, powdered husks, peat, and straw, can be added to reduce scabbing, hot tear, and hot crack casting defects when casting high temperature metals. These materials are beneficial because burn-off when the metal is poured creates tiny voids in the mold, allowing the sand particles to expand. They also increase collapsibility and reduce shakeout time.

Up to 2% of *cereal binders*, such as dextrin, starch, sulphite lye, and molasses, can be used to increase dry strength (the strength of the mold after curing) and improve surface finish. Cereal binders also improve collapsibility and reduce shakeout time because they burn off when the metal is poured. The disadvantage to cereal binders is that they are expensive.

Up to 2% of iron oxide powder can be used to prevent mold cracking and metal penetration, essentially improving refractoriness. Silica flour (fine silica) and zircon flour also improve refractoriness, especially in ferrous castings. The disadvantages to these additives is that they greatly reduce permeability.

Parting Compounds

To get the pattern out of the mold, prior to casting, a parting compound is applied to the pattern to ease removal. They can be a liquid or a fine powder (particle diameters between 75 and 150 micrometres (0.0030 and 0.0059 in)). Common powders include talc, graphite, and dry silica; common liquids include mineral oil and water-based silicon solutions. The latter are more commonly used with metal and large wooden patterns.

PERMANENT MOLD CASTING

Permanent mold casting is a metal casting process that employs reusable molds ("permanent molds"), usually made from metal. The most common process uses gravity to fill the mold, however gas pressure or a vacuum are also used. A variation on the typical gravity casting process, called slush casting, produces hollow castings. Common casting metals are aluminium, magnesium, and copper alloys. Other materials include tin, zinc, and lead alloys and iron and steel are also cast in graphite molds.

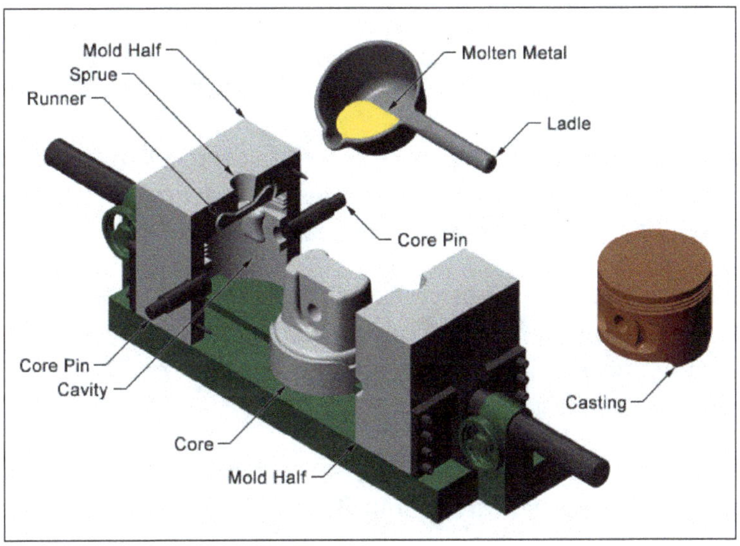

Permanent mold casting.

Typical products are components such as gears, splines, wheels, gear housings, pipe fittings, fuel injection housings, and automotive engine pistons.

Process

There are four main types of permanent mold casting: gravity, slush, low-pressure, and vacuum.

Gravity Process

The gravity process begins by preheating the mold to 150–200 °C (300–400 °F) to ease the flow and reduce thermal damage to the casting. The mold cavity is then coated with a refractory material or a mold wash, which prevents the casting from sticking to the mold and prolongs the mold life. Any sand or metal cores are then installed and the mold is clamped shut. Molten metal is then poured into the mold. Soon after solidification the mold is opened and the casting removed to reduce chances of hot tears. The process is then started all over again, but preheating is not required because the heat from the previous casting is adequate and the refractory coating should last several castings. Because this process is usually carried out on large production run work-pieces automated equipment is used to coat the mold, pour the metal, and remove the casting.

The metal is poured at the lowest practical temperature in order to minimize cracks and porosity. The pouring temperature can range greatly depending on the casting material; for instance zinc alloys are poured at approximately 370 °C (698 °F), while Gray iron is poured at approximately 1,370 °C (2,500 °F).

Mold

Molds for the casting process consist of two halves. Casting molds are usually formed from gray cast iron because it has about the best thermal fatigue resistance, but other materials include steel, bronze, and graphite. These metals are chosen because of their resistance to erosion and thermal fatigue. They are usually not very complex because the mold offers no collapsibility to compensate for shrinkage. Instead the mold is opened as soon as the casting is solidified, which prevents hot tears. Cores can be used and are usually made from sand or metal.

The mold is heated prior to the first casting cycle and then used continuously in order to maintain as uniform a temperature as possible during the cycles. This decreases thermal fatigue, facilitates metal flow, and helps control the cooling rate of the casting metal.

Venting usually occurs through the slight crack between the two mold halves, but if this is not enough then very small vent holes are used. They are small enough to let the air escape but not the molten metal. A riser must also be included to compensate for shrinkage. This usually limits the yield to less than 60%.

Mechanical ejectors in the form of pins are used when coatings are not enough to remove casts from the molds. These pins are placed throughout the mold and usually leave small round impressions on the casting.

Slush

Slush casting is a variant of permanent molding casting to create a *hollow casting* or *hollow cast*. In the process the material is poured into the mold and allowed to cool until a shell of material forms in the mold. The remaining liquid is then poured out to leave a hollow shell. The resulting casting has good surface detail but the wall thickness can vary. The process is usually used to cast ornamental products, such as candlesticks, lamp bases, and statuary, from low-melting-point materials. A similar technique is used to make hollow chocolate figures for Easter and Christmas.

The method was developed by William Britain in 1893 for the production of lead toy soldiers. It uses less material than solid casting, and results in a lighter and less expensive product. Hollow cast figures generally have a small hole where the excess liquid was poured out.

Similarly, a process called *slush molding* is used in automotive dashboard manufacture, for soft-panel interiors with artificial leather, where a free-flowing (which behave like a liquid) powder plastic compound, either PVC or TPU, is poured into a hot, hollow mold and a viscous skin forms. Excess slush is then drained off, the mold is cooled, and the molded product is stripped out.

Low-pressure

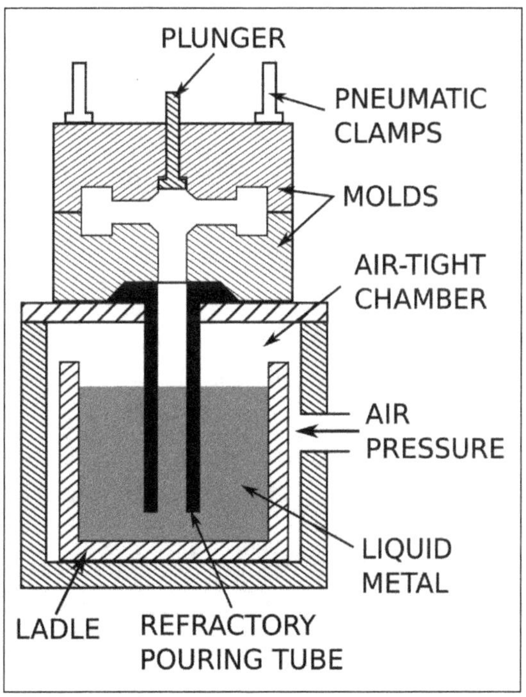

Schematic of the low-pressure permanent mold casting process.

Low-pressure permanent mold (*LPPM*) casting uses a gas at low pressure, usually between 3 and 15 psi (20 to 100 kPa) to push the molten metal into the mold cavity. The pressure is applied to the top of the pool of liquid, which forces the molten metal up a refractory pouring tube and finally into the bottom of the mold. The pouring tube extends to the bottom of the ladle so that the material being pushed into the mold is exceptionally clean. No risers are required because the applied pressure forces molten metal in to compensate for shrinkage. Yields are usually greater than 85% because there is no riser and any metal in the pouring tube just falls back into the ladle for reuse.

The vast majority of LPPM casting are from aluminum and magnesium, but some are copper alloys. Advantages include very little turbulence when filling the mold because of the constant pressure, which minimizes gas porosity and dross formation. Mechanical properties are about 5% better than gravity permanent mold castings. The disadvantage is that cycles times are longer than gravity permanent mold castings.

Vacuum

Vacuum permanent mold casting retains all of the advantages of LPPM casting, plus the dissolved gases in the molten metal are minimized and molten metal cleanliness is even better. The process can handle thin-walled profiles and gives an excellent surface finish. Mechanical properties are usually 10 to 15% better than gravity permanent mold castings. The process is limited in weight to 0.2 to 5 kg (0.44 to 11.02 lb).

Advantages and Disadvantages

The main advantages are the reusable mold, good surface finish, good dimensional accuracy, and high production rates. Typical tolerances are 0.4 mm for the first 25 mm (0.015 in for the first inch) and 0.02 mm for each additional centimeter (0.002 in per in); if the dimension crosses the parting line add an additional 0.25 mm (0.0098 in). Typical surface finishes are 2.5 to 7.5 μm (100–250 μin) RMS. A draft of 2 to 3° is required. Wall thicknesses are limited to 3 to 50 mm (0.12 to 1.97 in). Typical part sizes range from 100 g to 75 kg (several ounces to 150 lb). Other advantages include the ease of inducing directional solidification by changing the mold wall thickness or by heating or cooling portions of the mold. The fast cooling rates created by using a metal mold results in a finer grain structure than sand casting. Retractable metal cores can be used to create undercuts while maintaining a quick action mold.

There are three main disadvantages: high tooling cost, limited to low-melting-point metals, and short mold life. The high tooling costs make this process uneconomical for small production runs. When the process is used to cast steel or iron the mold life is extremely short. For lower melting point metals the mold life is longer but thermal fatigue and erosion usually limit the life to 10,000 to 120,000 cycles. The mold life is dependent on four factors: the mold material, the pouring temperature, the mold temperature, and the mold configuration. Molds made from gray cast iron can be more economical to produce but have short mold lives. On the other hand, molds made from H13 tool steel may have a mold life several times greater. The pouring temperature is dependent on the casting metal, but the higher the pouring temperature the shorter the mold life. A high pouring temperature can also induce shrinkage problems and create longer cycle times. If the mold temperature is too low misruns are produced, but if the mold temperature is too high then the cycle time is prolonged and mold erosion is increased. Large differences in section thickness in the mold or casting can decrease mold life as well.

RESIN CASTING

Resin casting is a method of plastic casting where a mold is filled with a liquid synthetic resin, which then hardens. It is primarily used for small-scale production like industrial prototypes and dentistry. It can be done by amateur hobbyists with little initial investment, and is used in the production of collectible toys, models and figures, as well as small-scale jewellery production.

The synthetic resin for such processes is a monomer for making a plastic thermosetting polymer. During the setting process, the liquid monomer polymerizes into the polymer, thereby hardening into a solid.

Single-monomer resins may be used in the process, which form homopolymers (polymers containing only one type of polymer). In such uses, the "curing agent" mixed with the resin contains what is loosely referred to as a "catalyst," but which is more technically an initial source of free radicals (such as MEKP) to act as an initiator in a free-radical chemical chain reaction polymerization. Alternately, resin casting may be accomplished with a resin plus a nearly equal amount of a "hardener" liquid (as in many epoxy resin or polyester resin systems), which functionally contains a second polymer, for use in forming a final product plastic which is a copolymer. Copolymers contain two different alternating chemical entities in the final polymer molecule.

Process

Most commonly a thermosetting resin is used that polymerizes by mixing with a curing agent (polymerization catalyst) at room temperature and normal pressure. The resins are named by analogy with plant resins, but are synthetic monomers for making polymer plastics. The so-called synthetic resins used include polystyrene resin, polyurethane resin, epoxy resin, unsaturated polyester resin, acrylic resin and silicone resin.

Epoxy resin has a lower viscosity than polyurethane resin; polyester resin also shrinks markedly while curing. Acrylic resin, in particular the methyl methacrylate type of synthetic resin, produces acrylic glass (also called PMMA, Lucite, Plexiglass), which is not a glass but a plastic polymer that is transparent, and very hard. It is suitable for embedding objects (such as, for example, acrylic trophies), for display purposes. Styrene is a similar liquid monomer at room temperature, which will also polymerize into clear glass-like polystyrene plastic, with addition of a suitable catalyst.

A flexible mold can be made of latex rubber, room temperature vulcanized silicone rubber or other similar materials at relatively low cost, but can only be used for a limited number of castings.

The simplest method is gravity casting where the resin is poured into the mold and pulled down into all the parts by gravity. When the two part resin is mixed air bubbles tend to be introduced into the liquid which can be removed in a vacuum chamber. The casting can also be done in a vacuum chamber (when using open molds) to either extract these bubbles, or in a pressure pot, to reduce their size to the point where they aren't visible. Pressure and/or centrifugal force can be used to help push the liquid resin into all details of the mold. The mold can also be vibrated to expel bubbles.

Each unit requires some amount of hands-on labor, making the final cost per unit produced fairly high. This is in contrast to injection molding where the initial cost of creating the metal mold is higher, but the mold can be used to produce a much higher number of units, resulting in a lower cost per unit.

Collectibles and Models

Resin casting is used to produce collectible and customized toys and figures like designer toys, garage kits and ball-jointed dolls, as well as scale models, either individual parts or entire models of objects like trains, aircraft or ships. They are generally produced in small quantities, from the tens to a few hundred copies, compared to injection-molded plastic figures which are produced in many thousands. Resin casting is more labor intensive than injection molding, and the soft molds used are worn down by each cast. The low initial investment cost of resin casting means that individual hobbyists can produce small runs for their own use, such as customization, while companies can use it to produce small runs for public sale.

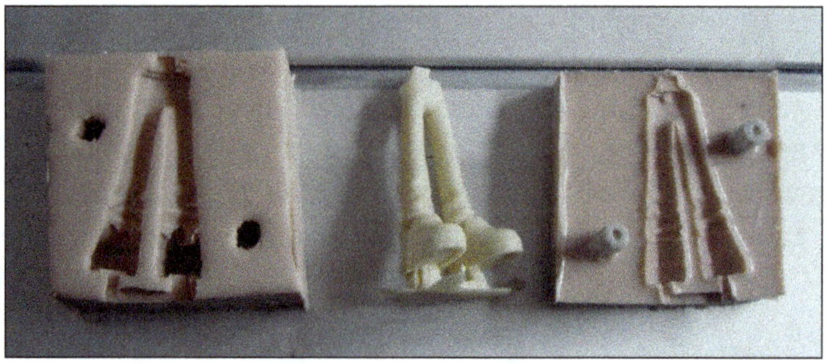

A custom resin cast Pinky:St part and two-part silicone mold.

The creation of a toy or figure start with the traditional sculpting process where the artist designs a clay sculpture. Where appropriate, for example when making a garage kit, the sculpture is dissected into several parts like head, torso, arms and legs. A flexible mold made from room temperature vulcanized (RTV) silicone rubber is made for each part.

After the mold has been made, a synthetic resin - such as polyurethane or epoxy - mixed with a curing agent, is poured into each mold cavity. Mixing the two liquid parts causes an exothermic reaction which generates heat and within minutes causes the material to harden, yielding castings or copies in the shape of the mold into which it has been poured. The molds are commonly half-divided (like the hollowed chocolate Easter eggs with candy inside) and a release agent may be used to make removal of the hardened/set resin from the mold easier. The hardened resin casting is removed from the flexible mold and allowed to cool.

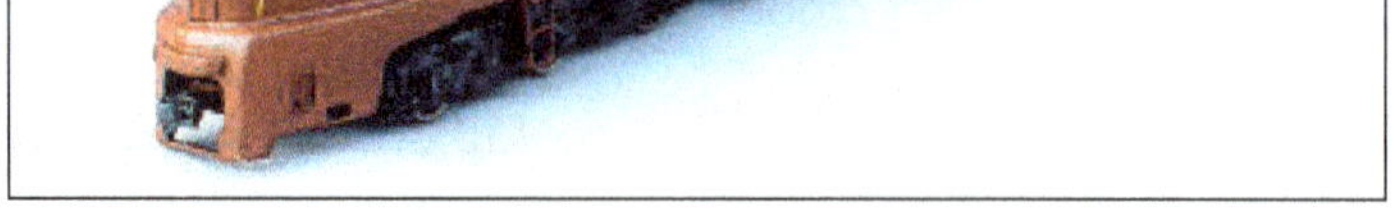

A Baldwin 6-axle locomotive kit cast in resin in HO Scale.

Due to aggressive nature of most compounds used for casting and the high temperature of the reaction the mold gradually degrades and loses small details. Typically, a flexible mold will yield between 25 and 100 castings depending upon the size of the part, the intensity of the heat generated.

Depending on the type of product it may then be cut or sanded to remove any casting artefacts like sprues and seams. Some products are also assembled and painted, while some models and kits, which are intended for the consumer to assemble, are left unfinished.

The ability of RTV silicone molds to reproduce even the tiniest detail means that many of these low volume castings are of very high quality. Quality of both original masters and resin castings varies due to differences in creator's skill, as well as casting techniques.

SPRAY FORMING

Spray forming, also known as spray casting, spray deposition and in-situ compaction, is a method of casting near net shape metal components with homogeneous microstructures via the deposition of semi-solid sprayed droplets onto a shaped substrate. In spray forming an alloy is melted, normally in an induction furnace, then the molten metal is slowly poured through a conical tundish into a small-bore ceramic nozzle. The molten metal exits the furnace as a thin free-falling stream and is broken up into droplets by an annular array of gas jets, and these droplets then proceed downwards, accelerated by the gas jets to impact onto a substrate. The process is arranged such that the droplets strike the substrate whilst in the semi-solid condition, this provides sufficient liquid fraction to 'stick' the solid fraction together. Deposition continues, gradually building up a spray formed billet of metal on the substrate.

The *gas atomised spray forming* (*GASF*) process typically has a molten alloy flow rate of 1–20 kg/min, although twin atomizer systems can achieve metal flow rates of up to 80 kg/min. Special steel billets of 1 tonne or more have been produced by spray forming on a commercial basis, together with Ni super-alloy ring blanks of up to 500 kg and Al alloy extrusion billets of up to 400 kg.

Advantages

Spray forming offers certain advantages over both conventional ingot metallurgy and more specialized techniques such as powder metallurgy. Firstly, it is a flexible process and can be used to manufacture a wide range of materials, some of which are difficult to produce by other methods, e.g. Al-5wt% Li alloys or Al-SiC, Al-Al_2O_3 metal matrix composites (MMCs). The atomisation of the melt stream into droplets of 10-500 μm diameter, some of which, depending on diameter, cool quickly to the solid and semi-solid state provide a large number of nucleants for the residual liquid fraction of the spray formed material on the billet top surface. The combination of rapid cooling in the spray and the generation of a large population of solid nucleants in the impacting spray leads to a fine equiaxed microstructure, typically in the range 10–100 μm, with low levels and short length scales of internal solute partitioning. These microstructural aspects offer advantages in material strength because of fine grain size, refined distribution of dispersoid and/or secondary precipitate phases, as well as tolerance to impurity 'tramp' elements. This fine structure in the 'as sprayed' condition means homogenising heat treatments can often be avoided. Because of the complex solidification path (i.e. the rapid transition from superheated melt to solid, liquid or semi-solid droplet to temperature equilibration at semi-solid billet top and final slow cooling to fully solid) of the spray formed material, extended solubility of alloying elements and the formation of metastable and quasi-crystalline phases has also been reported.

One of the major attractions of spray forming is the potential economic benefit to be gained from reducing the number of process steps between melt and finished product. Spray forming can be used to produce strip, tube, ring, clad bar/roll and cylindrical extrusion feed stock products, in each case with a relatively fine-scale microstructure even in large cross-sections. The benefits of GASF over powder metallurgy accrue from the reduced number of process steps where powder sieving, pressing, de-gassing and handling steps and their attendant safety and contamination issues may be removed.

Disadvantages

There are two major disadvantages to the gas atomisation spray forming process. The most significant disadvantage is the relatively low process yield with typical losses of ~30%. Losses occur because of overspray (droplets missing the emerging billet), splashing of material from the billet surface, and material 'bouncing' off the semi-solid top surface. Many operators of the spray forming process now use a particle injector system to re-inject the overspray powder, and thus recycle material that would otherwise be lost, or sell the overspray powder as a product in its own right. The second major disadvantage is one of process control. As it is essentially a free-forming process with many interdependent variables, it has proved difficult to predict the shape, porosity or deposition rate for a given alloy. Much of the control is based on operator experience and empirical relationships. It is partly the process complexity and lack of robust process control that has prevented the widespread commercialisation of this process. Some developments using feed-back control have proved successful in improving the variations in billet diameter and improving yield in specific systems but these have yet to find widespread implementation.

Porosity resulting from gas entrapment and solidification shrinkage is a significant problem in spray formed materials. A typical spray formed billet will contain 1-2% porosity with a pore size dependent on alloy freezing range and various process parameters. Hot isostatic pressing (HIPing) or thermo-mechanical processing can heal these pores if they are small (less than 30 μm). Despite these disadvantages, spray forming remains an economic process for the production of difficult to manufacture, niche alloys. Large-scale porosity is more difficult to heal effectively and must be minimised by careful process control. In some cases, porosity is controlled by alloy additions which react with dissolved and entrapped gas to form a solid phase e.g. titanium added to copper billets to form titanium nitride with dissolved and entrapped nitrogen gas. Porosity, even after consolidation, can limit the applications of spray formed material, for example rotating gas turbine components must have zero porosity because of the detrimental effect on high-cycle fatigue (HCF).

Commercialisation

In spite of the problems associated with the spray forming process there has been sustained industrial interest in spray forming over the last 35 years. Sandvik-Osprey of Neath, South Wales holds the patents on the process and have licensed the technology to a range of industries. There are currently approximately 25 licensees operating around the world, ranging from small research and development plants to full-scale commercial operations. Main applications are prematerial for low temperature Nb_3Sn super conductors (CuSn), oil drilling equipment (high strength material CuMnNi) and for forming tools (CuAlFe with high Al-content). In all of these applications, research concerns the reconciliation of the cost disadvantages and complexity of spray forming with the demand for high performance alloys in niche applications.

Melting

The earliest spray forming work was based on a resistively heated electric holding furnace. The melt then passed through a 3 mm diameter Al_2O_3 nozzle. However the low flow rate made a high superheat necessary to prevent solidification in the nozzle. The next generation melting procedures in spray forming applications were bottom pour induction units, which offer many benefits. In this system, the melting crucible is directly above the atomiser head with a ceramic nozzle feeding directly from the furnace to the atomiser. A stopper rod runs through the melt to the top of the pouring nozzle, the rod is withdrawn when the melt reaches the designated temperature for spraying, typically 50 to 150 °C (122 to 302 °F) above the alloy's liquidus. Alternatively a pre-prepared plug of alloy to block the nozzle is used, and at a specified superheat this plug melts allowing the contents of the furnace to drain through the nozzle. Another problem associated with bottom pour furnaces is the change in flow rate associated with the reducing metalo-static head in the crucible. In some cases, introducing an inert gas overpressure during pouring can compensate for this effect.

An alternative approach is the tilt-pour furnace whereby an induction furnace is tilted to pour the melt into a conical tundish that in turn delivers the molten metal to the melt delivery nozzle. The tilt pour system provides the advantage that melting is decoupled from the spraying procedure so that melting problems and remedial solutions do not affect or disturb the critical set-up of the melt delivery nozzle.

In the most complex melting arrangement, used only for production of nickel superalloy turbine forging blanks by spray forming, vacuum induction melting, electroslag re-melting and cold hearth crucibles have been combined by GE to control alloy impurity levels and the presence of refractory inclusions in the molten metal supply. *Clean metal spray forming* (*CMSF*) combines the electroslag refining process, cold walled induction guide and gas atomised spray forming. This approach has led to a reduction in the number of melt related defects (pores, inclusions, etc.), a finer average grain size, the ability to produce larger ingots and the ability to process a wider range of alloys.

Atomisation

There are many different techniques for atomisation of molten metals, many of which are derived from the powder metallurgy industry and have been extensively reviewed elsewhere. There are two major atomisation techniques used in spray forming: centrifugal atomisation for the manufacture of near net shape rings and gas atomisation for the manufacture of billets, tube and strip.

Centrifugal Atomisation

Centrifugal atomisation involves pouring molten metal at relatively low flow rates (0.1– 2 kg/min) onto a spinning plate, dish or disc, whereby the rotation speed is sufficient to create high centrifugal forces at the periphery and overcome surface tension and viscous forces so the melt is fragmented into droplets. Droplet diameters produced by centrifugal atomisation are dependent primarily on the rotation speed, (up to 20,000 rpm) and are typically in the range 20–1000 μm with cooling rates of the order 104 Ks^{-1}. Centrifugal atomisation is generally conducted under an inert atmosphere of Ar or N_2 to prevent oxidation of the fine droplets or can be operated under vacuum.

Gas Atomisation

The melt stream exits the melt delivery nozzle into the spray chamber. The melt stream is protected from being destabilised by the turbulent gas environment in the spray chamber by primary gas jets operating at intermediate inert gas pressure of 2 to 4 bar, the resulting gas flow is parallel to the melt stream to stabilise the melt stream. The secondary atomiser uses high velocity (250 to 350 ms^{-1}), high-pressure (6 to 10 bar) gas jets to impinge on the melt stream to achieve atomisation. The atomiser jets are usually arranged as an annulus or as discrete jets positioned symmetrically about the melt delivery nozzle, or less commonly, arranged as a linear nozzle for the production of strip products. Typical droplet diameters follow a log-normal distribution with powder diameters up to ~600 μm with a mass median diameter of ~150 μm.

The atomising gas mass flow rate to molten metal mass flow rate ratio is a key parameter in controlling the droplet diameter and hence the cooling rate, billet temperature and resulting solid particle nucleant density. The gas-metal ratio (GMR) is typically in the range 1.5 to 5.5, with yield decreasing and cooling rates in the spray increasing with increasing GMR. Typically at low (1.5) GMR, yield is 75%, if the GMR is increased to 5.0 with all other parameters remaining constant, the process yield is reduced to 60%.

Scanning atomisers have been developed which allow the production of billets of up to 600 mm diameter, approximately twice the diameter possible with a static atomiser. The atomiser head is oscillated mechanically through 5 to 10° at a typical frequency of 25 Hz, to deflect the melt stream creating a spray path that is synchronised with the rotation speed of the collector plate in order to deposit a parallel-sided billet. By using programmable oscillating atomiser drives it was possible to improve the shape and shape reproducibility of spray formed deposits. It has been demonstrated that parallel sided, flat topped billets could be sprayed in a reproducible manner if the substrate rotation and atomiser oscillation frequency were synchronised and optimised for specific alloys and melt flow rates. Twin atomiser systems combine a static and scanning atomiser, making it possible to spray billets of up to 450 mm diameter with economic benefits.

Atomising gas used in spray forming is generally either N_2 and can be either protective or reactive depending on the alloy system, or Ar which is generally entirely inert but more expensive than N_2. Reactive gasses can be introduced in small quantities to the atomising gas to create dispersion strengthened alloys e.g. 0.5–10% O_2 in N_2 used to generate oxide dispersion strengthened (ODS) Al alloys. Comparisons of N_2 and Ar based spray forming showed that with all other factors remaining constant, the billet top temperature was lower with N_2 than with Ar, because of the differences in thermal diffusivity of the two atomising gases: Ar has a thermal conductivity of 0.0179 W/mK which is approximately a third less than N_2 with a thermal conductivity of 0.026 W/mK.

The mechanisms of melt break up and atomisation have been extensively researched, showing that atomisation typically consists of 3 steps: (1) primary break up of the melt stream; (2) molten droplets and ligaments undergo secondary disintegration; (3) particles cool and solidify. Theoretical analysis of the atomisation process to predict droplet size has yielded models providing only moderate agreement with experimental data.

Investigations show that in all cases gas atomisation of molten metal yields a broad range of droplet diameters, typically in the range 10-600 µm diameter, with a mean diameter of ~100 µm. Droplet diameter governs the dynamic behaviour of the droplet in flight which in turn determines the time available for in-flight cooling which is critical in controlling the resulting billet microstructure. At a flight distance of 300–400 mm, predictions show droplet velocities of 40-90 ms^{-1} for droplet diameters in the range 20-150 µm respectively, compared to measured velocities of ~100 ms^{-1}, and at distances of up to 180 mm from the atomiser, droplets were still being accelerated by the gas. Droplets cool in-flight predominantly by convection and radiation, and can experience undercooling of up to 300 °C (572 °F) prior to nucleation. Models and experimental measurements show that small droplets (<50 µm) very rapidly become fully solid prior to deposition, 50-200 µm droplets will be typically semi-solid and droplets of diameters >200 µm will be liquid at deposition. The range of droplet dynamic and thermal histories result in a billet top surface of 0.3 to 0.6 solid fraction. Not all material that impacts the surface is incorporated into the billet: some solid droplets will bounce or splash-off the billet top surface or be directed out of the deposition region by turbulent gas movement in the chamber. The proportion of droplets that impact the surface compared to the proportion that are incorporated into the billet has been termed the *sticking efficiency*: dependent on the geometric sticking which is a function of the spray angle relative to substrate and the thermal sticking efficiency dependent on spray and billet solid/liquid fraction.

Spray formed Microstructure

During spraying it is essential to maintain a constant top surface temperature and hence maintain steady-state conditions if a billet with consistent microstructure is to be produced. At the billet surface, during spraying an enthalpy balance must be maintained where the rate of enthalpy lost (H_{out}) from the billet by conduction to the atomising gas and through the substrate, convection and radiation must be balanced with the rate of enthalpy input (H_{in}) from the droplets in the spray. There are a variety of factors that can be adjusted in order to maintain these conditions: spray height, atomiser gas pressure, melt flow rate, melt superheat and atomiser configuration, being those parameters most readily adjusted. Typically equipment such as closed circuit cameras and optical pyrometry can be used to monitor billet size/position and top surface temperature. If H_{out} is much greater H_{in} then a steady temperature is maintained at the billet top surface. The top surface should be in a mushy condition in order to promote sticking of incoming droplets and partial re-melting of solid particles. The necessary partial re-melting of solid droplets explains the absence of dendritic remnants from pre-solidified droplets in the final microstructure. If Hin is insufficient to cause significant re-melting, a 'splat' microstructure of layered droplets will form, typical of thermal spray processes such as vacuum plasma spraying (VPS), arc spraying and high velocity oxy-fuel. Processing maps have been produced for plasma spraying and spray forming using a steady-state heat balance in terms of the interlayer time (time between deposition events) against average deposition rate per unit area. These maps show the boundaries between banded un-fused microstructure and an equiaxed homogeneous structure.

The final phase of solidification occurs once droplets have impacted the mushy billet surface and thermal equilibration has taken place between the droplets and the billet. At this stage residual liquid is present as continuous network delineating polygonal grain boundaries, with a typical liquid fraction of 0.3 – 0.5. The cooling rates during solidification of the billet is several orders of magnitude slower than the cooling rate in the spray, at 1-20 Ks^{-1}.

Although one of the benefits of spray forming is purportedly the ability to produce bulk material with fine scale microsegregation and little or no macrosegregation work on Al-Mg-Li-Cu alloys showed that as a consequence of the interconnected liquid in the billet there was significant macrosegregation in large spray formed wrought Al billets. The distribution of Cu, Mg and Li in, for example, Al alloy 8091 showed surprisingly pronounced macrosegregation with the variation of Cu(wt%) in a spray formed 8091 billet, ranging from approximately 1.4 at the billet centre to 1.92 at the billet periphery. These macrosegregation patterns were explained in terms of inverse segregation in which solute rich liquid from the billet centre is sucked back through the primary Al-rich network to feed solidification shrinkage at the billet periphery. This effect was suggested to be exacerbated by centrifugal effects from the billet rotation.

As sprayed the billet porosity is typically 1–2% with a region of higher porosity in the splat-quenched region adjacent to the substrate. The very top of the billet often shows increased porosity because the top is rapidly chilled by the atomising gas which continues to chill the billet for 10–60 seconds after spraying. There has also been little progress in understanding and quantifying the underlying physics that controls as-sprayed porosity.

In most cases, the higher porosity at the billet base and top are scalped and recycled. Ultrasonic inspection is sometimes used to determine the depth of the chill zone regions to prevent unnecessary wastage. Depending on the alloy system and the final application, the remaining bulk material is usually processed to close porosity and subjected to a range of thermo-mechanical treatments. Spray formed materials are rarely used in the as-sprayed condition and are often treated by HIPing to remove porosity. In some cases, the residual atomising gas in pores may react with alloying elements to form allegedly beneficial phases e.g. N_2 reacting with titanium in nickel superalloy Rene 80 to form a dispersion of TiN.

DIE CASTING

Die casting is a metal casting process that is characterized by forcing molten metal under high pressure into a mold cavity. The mold cavity is created using two hardened tool steel dies which have been machined into shape and work similarly to an injection mold during the process. Most die castings are made from non-ferrous metals, specifically zinc, copper, aluminium, magnesium, lead, pewter, and tin-based alloys. Depending on the type of metal being cast, a hot- or cold-chamber machine is used.

An engine block with aluminium and magnesium die castings.

The casting equipment and the metal dies represent large capital costs and this tends to limit the process to high-volume production. Manufacture of parts using die casting is relatively simple, involving only four main steps, which keeps the incremental cost per item low. It is especially suited for a large quantity of small- to medium-sized castings, which is why die casting produces more castings than any other casting process. Die castings are characterized by a very good surface finish (by casting standards) and dimensional consistency.

Cast Metal

The main die casting alloys are: zinc, aluminium, magnesium, copper, lead, and tin; although uncommon, ferrous die casting is also possible. Specific die casting alloys include: zinc aluminium; aluminium to, e.g. The Aluminum Association (AA) standards: AA 380, AA 384, AA 386, AA 390; and AZ91D magnesium. The following is a summary of the advantages of each alloy:

- Zinc: The easiest metal to cast; high ductility; high impact strength; easily plated; economical for small parts; promotes long die life.
- Aluminium: Lightweight; high dimensional stability for complex shapes and thin walls; good corrosion resistance; good mechanical properties; high thermal and electrical conductivity; retains strength at high temperatures.
- Magnesium: The easiest metal to machine; excellent strength-to-weight ratio; lightest alloy commonly die cast.
- Copper: High hardness; high corrosion resistance; highest mechanical properties of alloys die cast; excellent wear resistance; excellent dimensional stability; strength approaching that of steel parts.
- Silicon tombac: High-strength alloy made of copper, zinc and silicon. Often used as an alternative for investment cast steel parts.
- Lead and tin: High density; extremely close dimensional accuracy; used for special forms of corrosion resistance. Such alloys are not used in foodservice applications for public health reasons. Type metal, an alloy of lead, tin and antimony (with sometimes traces of copper) is used for casting hand-set type in letterpress printing and hot foil blocking. Traditionally cast in hand jerk molds now predominantly die cast after the industrialisation of the type foundries. Around 1900 the slug casting machines came onto the market and added further automation, with sometimes dozens of casting machines at one newspaper office.

Maximum weight limits for aluminium, brass, magnesium, and zinc castings are approximately 70 pounds (32 kg), 10 lb (4.5 kg), 44 lb (20 kg), and 75 lb (34 kg), respectively.

The material used defines the minimum section thickness and minimum draft required for a casting as outlined in the table below. The thickest section should be less than 13 mm (0.5 in), but can be greater.

Metal	Minimum section	Minimum draft
Aluminium alloys	0.89 mm (0.035 in)	1:100 (0.6°)
Brass and bronze	1.27 mm (0.050 in)	1:80 (0.7°)

Magnesium alloys	1.27 mm (0.050 in)	1:100 (0.6°)
Zinc alloys	0.63 mm (0.025 in)	1:200 (0.3°)

Design Geometry

There are a number of geometric features to be considered when creating a parametric model of a die casting:

- Draft is the amount of slope or taper given to cores or other parts of the die cavity to allow for easy ejection of the casting from the die. All die cast surfaces that are parallel to the opening direction of the die require draft for the proper ejection of the casting from the die. Die castings that feature proper draft are easier to remove from the die and result in high-quality surfaces and more precise finished product.

- Fillet is the curved juncture of two surfaces that would have otherwise met at a sharp corner or edge. Simply, fillets can be added to a die casting to remove undesirable edges and corners.

- Parting line represents the point at which two different sides of a mold come together. The location of the parting line defines which side of the die is the cover and which is the ejector.

- Bosses are added to die castings to serve as stand-offs and mounting points for parts that will need to be mounted. For maximum integrity and strength of the die casting, bosses must have universal wall thickness.

- Ribs are added to a die casting to provide added support for designs that require maximum strength without increased wall thickness.

- Holes and windows require special consideration when die casting because the perimeters of these features will grip to the die steel during solidification. To counteract this effect, generous draft should be added to hole and window features.

Equipment

There are two basic types of die casting machines: *hot-chamber machines* and *cold-chamber machines*. These are rated by how much clamping force they can apply. Typical ratings are between 400 and 4,000 st (2,500 and 25,400 kg).

Hot-chamber Die Casting

Hot-chamber die casting, also known *gooseneck machines*, rely upon a pool of molten metal to feed the die. At the beginning of the cycle the piston of the machine is retracted, which allows the molten metal to fill the "gooseneck". The pneumatic- or hydraulic-powered piston then forces this metal out of the gooseneck into the die. The advantages of this system include fast cycle times (approximately 15 cycles a minute) and the convenience of melting the metal in the casting machine. The disadvantages of this system are that it is limited to use with low-melting point metals and that aluminium cannot be used because it picks up some of the iron while in the molten pool. Therefore, hot-chamber machines are primarily used with zinc-, tin-, and lead-based alloys.

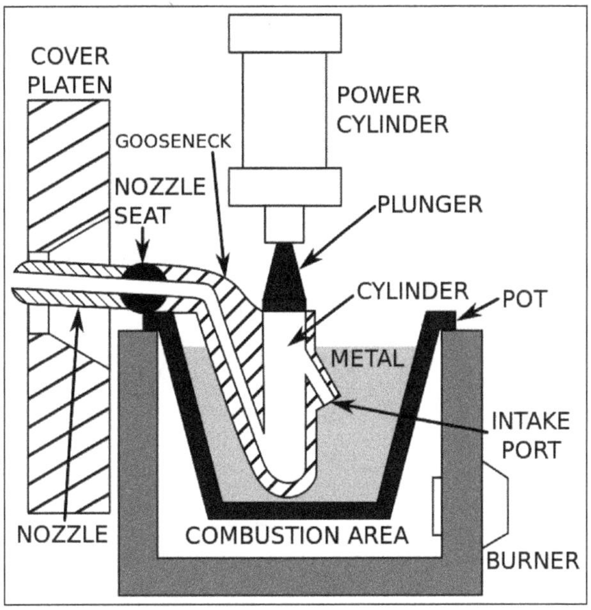

Schematic of a hot-chamber machine.

Cold-chamber Die Casting

These are used when the casting alloy cannot be used in hot-chamber machines; these include aluminium, zinc alloys with a large composition of aluminium, magnesium and copper. The process for these machines start with melting the metal in a separate furnace. Then a precise amount of molten metal is transported to the cold-chamber machine where it is fed into an unheated shot chamber (or injection cylinder). This shot is then driven into the die by a hydraulic or mechanical piston. The biggest disadvantage of this system is the slower cycle time due to the need to transfer the molten metal from the furnace to the cold-chamber machine.

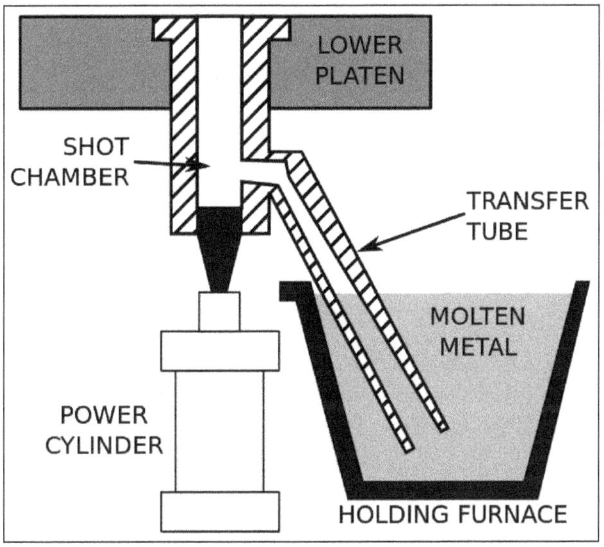

A schematic of a cold-chamber die casting machine.

Complete working cell.

Mold or Tooling

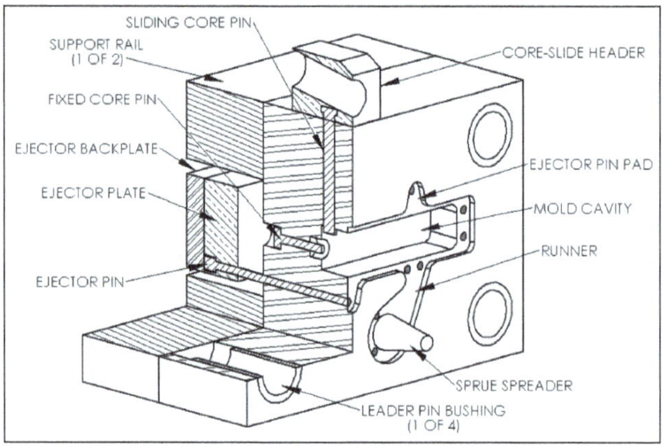

The ejector die half.

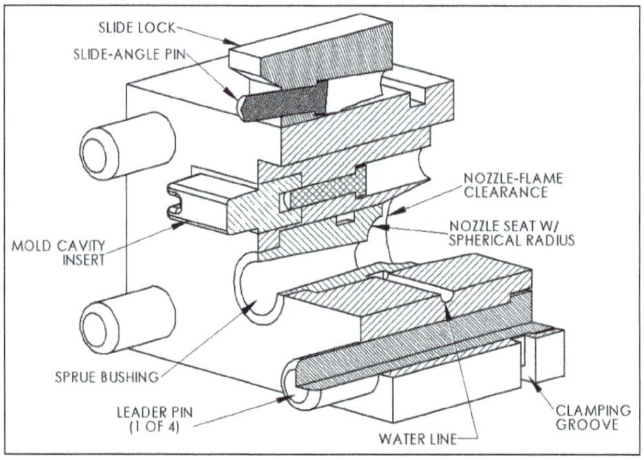

The cover die half.

Two dies are used in die casting; one is called the "cover die half" and the other the "ejector die half". Where they meet is called the parting line. The cover die contains the sprue (for hot-chamber machines) or shot hole (for cold-chamber machines), which allows the molten metal to flow into the dies; this feature matches up with the injector nozzle on the hot-chamber machines or the shot chamber in the cold-chamber machines. The ejector die contains the ejector pins and usually the runner, which is the path from the sprue or shot hole to the mold cavity. The cover die is secured to the stationary, or front, platen of the casting machine, while the ejector die is attached to the movable platen. The mold cavity is cut into two *cavity inserts*, which are separate pieces that can be replaced relatively easily and bolt into the die halves.

The dies are designed so that the finished casting will slide off the cover half of the die and stay in the ejector half as the dies are opened. This assures that the casting will be ejected every cycle because the ejector half contains the *ejector pins* to push the casting out of that die half. The ejector pins are driven by an *ejector pin plate*, which accurately drives all of the pins at the same time and with the same force, so that the casting is not damaged. The ejector pin plate also retracts the pins after ejecting the casting to prepare for the next shot. There must be enough ejector pins to keep the overall force on each pin low, because the casting is still hot and can be damaged by excessive force. The pins still leave a mark, so they must be located in places where these marks will not hamper the casting's purpose.

Other die components include *cores* and *slides*. Cores are components that usually produce holes or opening, but they can be used to create other details as well. There are three types of cores: fixed, movable, and loose. Fixed cores are ones that are oriented parallel to the pull direction of the dies (i.e. the direction the dies open), therefore they are fixed, or permanently attached to the die. Movable cores are ones that are oriented in any other way than parallel to the pull direction. These cores must be removed from the die cavity after the shot solidifies, but before the dies open, using a separate mechanism. Slides are similar to movable cores, except they are used to form undercut surfaces. The use of movable cores and slides greatly increases the cost of the dies. Loose cores, also called *pick-outs*, are used to cast intricate features, such as threaded holes. These loose cores are inserted into the die by hand before each cycle and then ejected with the part at the end of the cycle. The core then must be removed by hand. Loose cores are the most expensive type of core, because of the extra labor and increased cycle time. Other features in the dies include water-cooling passages and vents along the parting lines. These vents are usually wide and thin (approximately 0.13 mm or 0.005 in) so that when the molten metal starts filling them the metal quickly solidifies and minimizes scrap. No risers are used because the high pressure ensures a continuous feed of metal from the gate.

The most important material properties for the dies are thermal shock resistance and softening at elevated temperature; other important properties include hardenability, machinability, heat checking resistance, weldability, availability (especially for larger dies), and cost. The longevity of a die is directly dependent on the temperature of the molten metal and the cycle time. The dies used in die casting are usually made out of hardened tool steels, because cast iron cannot withstand the high pressures involved, therefore the dies are very expensive, resulting in high start-up costs. Metals that are cast at higher temperatures require dies made from higher alloy steels.

Table: Die and component material and hardness for various cast metals.

Die component	Cast metal					
	Tin, lead & zinc		Aluminium & magnesium		Copper & brass	
	Material	Hardness	Material	Hardness	Material	Hardness
Cavity inserts	P20	290–330 HB	H13	42–48 HRC	DIN 1.2367	38–44 HRC
	H11	46–50 HRC	H11	42–48 HRC	H20, H21, H22	44–48 HRC
	H13	46–50 HRC				

Cores	H13	46–52 HRC	H13	44–48 HRC	DIN 1.2367	40–46 HRC
			DIN 1.2367	42–48 HRC		
Core pins	H13	48–52 HRC	DIN 1.2367 prehard	37–40 HRC	DIN 1.2367 prehard	37–40 HRC
Sprue parts	H13	48–52 HRC	H13 DIN 1.2367	46–48 HRC 44–46 HRC	DIN 1.2367	42–46 HRC
Nozzle	420	40–44 HRC	H13	42–48 HRC	DIN 1.2367 H13	40–44 HRC 42–48 HRC
Ejector pins	H13	46–50 HRC	H13	46–50 HRC	H13	46–50 HRC
Plunger shot sleeve	H13	46–50 HRC	H13 DIN 1.2367	42–48 HRC 42–48 HRC	DIN 1.2367 H13	42–46 HRC 42–46 HRC
Holder block	4140 prehard	~300 HB	4140 prehard	~300 HB	4140 prehard	~300 HB

The main failure mode for die casting dies is wear or erosion. Other failure modes are *heat checking* and *thermal fatigue*. Heat checking is when surface cracks occur on the die due to a large temperature change on every cycle. Thermal fatigue is when surface cracks occur on the die due to a large number of cycles.

Table: Typical die temperatures and life for various cast materials.

	Zinc	Aluminium	Magnesium	Brass (leaded yellow)
Maximum die life [number of cycles]	1,000,000	100,000	100,000	10,000
Die temperature [C° (F°)]	218 (425)	288 (550)	260 (500)	500 (950)
Casting temperature [C° (F°)]	400 (760)	660 (1220)	760 (1400)	1090 (2000)

Process

The following are the four steps in *traditional die casting*, also known as *high-pressure die casting*, these are also the basis for any of the die casting variations: die preparation, filling, ejection, and shakeout. The dies are prepared by spraying the mold cavity with lubricant. The lubricant both helps control the temperature of the die and it also assists in the removal of the casting. The dies are then closed and molten metal is injected into the dies under high pressure; between 10 and 175 megapascals (1,500 and 25,400 psi). Once the mold cavity is filled, the pressure is maintained until the casting solidifies. The dies are then opened and the shot (shots are different from castings because there can be multiple cavities in a die, yielding multiple castings per shot) is ejected by the ejector pins. Finally, the shakeout involves separating the scrap, which includes the gate, runners, sprues and flash, from the shot. This is often done using a special trim die in a power press or hydraulic press. Other methods of shaking out include sawing and grinding. A less labor-intensive method is to tumble shots if gates are thin and easily broken; separation of gates from finished parts must follow. This scrap is recycled by remelting it. The yield is approximately 67%.

The high-pressure injection leads to a quick fill of the die, which is required so the entire cavity fills before any part of the casting solidifies. In this way, discontinuities are avoided, even if the shape requires difficult-to-fill thin sections. This creates the problem of air entrapment, because when the mold is filled quickly there is little time for the air to escape. This problem is minimized by including vents along the parting lines, however, even in a highly refined process there will still be some porosity in the center of the casting.

Most die casters perform other secondary operations to produce features not readily castable, such as tapping a hole, polishing, plating, buffing, or painting.

Inspection

After the shakeout of the casting it is inspected for defects. The most common defects are misruns and cold shuts. These defects can be caused by cold dies, low metal temperature, dirty metal, lack of venting, or too much lubricant. Other possible defects are gas porosity, shrinkage porosity, hot tears, and flow marks. *Flow marks* are marks left on the surface of the casting due to poor gating, sharp corners, or excessive lubricant.

Lubricants

Water-based lubricants, called *emulsions*, are the most used type of lubricant, because of health, environmental, and safety reasons. Unlike solvent-based lubricants, if water is properly treated to remove all minerals from it, it will not leave any by-product in the dies. If the water is not properly treated, then the minerals can cause surface defects and discontinuities. There are four types of water-based lubricants: oil in water, water in oil, semi-synthetic, and synthetic. Oil in water is the best, because when the lubricant is applied, the water cools the die surface by evaporating while depositing the oil, which helps release the shot. A common mixture for this type of lubricants is thirty parts water to one part oil, however in extreme cases a ratio of one hundred to one is used.

Oils that are used include heavy residual oil (HRO), animal fats, vegetable fats, and synthetic fats. HROs are gelatinous at room temperature, but at the high temperatures found in die casting, they form a thin film. Other substances are added to control the emulsions viscosity and thermal properties; these include graphite, aluminium, and mica. Other chemical additives are used to inhibit rusting and oxidation. Emulsifiers are added to water-based lubricants, so that oil based additives can be mixed into the water; these include soap, alcohol esters, and ethylene oxides.

Historically, solvent-based lubricants, such as diesel fuel and kerosene, were commonly used. These were good at releasing the part from the dies, but a small explosion occurred during each shot, which led to a build-up of carbon on the mold cavity walls. However, they were easier to apply evenly than water-based lubricants.

Advantages and Disadvantages

Advantages of die casting:

- Excellent dimensional accuracy (dependent on casting material, but typically 0.1 mm for the first 2.5 cm (0.004 inch for the first inch) and 0.02 mm for each additional centimeter (0.002 inch for each additional inch).
- Smooth cast surfaces (Ra 1–2.5 micrometres or 0.04–0.10 thou rms).

- Thinner walls can be cast as compared to sand and permanent mold casting (approximately 0.75 mm or 0.030 in).
- Inserts can be cast-in (such as threaded inserts, heating elements, and high strength bearing surfaces).
- Reduces or eliminates secondary machining operations.
- Rapid production rates.
- Casting tensile strength as high as 415 megapascals (60 ksi).
- Casting of low fluidity metals.

The main disadvantage to die casting is the very high capital cost. Both the casting equipment required and the dies and related components are very costly, as compared to most other casting processes. Therefore, to make die casting an economic process, a large production volume is needed. Other disadvantages are that the process is limited to high-fluidity metals, and casting weights must be between 30 grams (1 oz) and 10 kg (20 lb). In the standard die casting process the final casting will have a small amount of porosity. This prevents any heat treating or welding, because the heat causes the gas in the pores to expand, which causes micro-cracks inside the part and exfoliation of the surface. Thus a related disadvantage of die casting is that it is only for parts in which softness is acceptable. Parts needing hardening (through hardening or case hardening) and tempering are not cast in dies.

Variants

Acurad

Acurad was a die casting process developed by General Motors in the late 1950s and 1960s. The name is an acronym for accurate, reliable, and dense. It was developed to combine a stable fill and directional solidification with the fast cycle times of the traditional die casting process. The process pioneered four breakthrough technologies for die casting: thermal analysis, flow and fill modeling, heat treatable and high integrity die castings, and indirect squeeze casting.

The thermal analysis was the first done for any casting process. This was done by creating an electrical analog of the thermal system. A cross-section of the dies were drawn on Teledeltos paper and then thermal loads and cooling patterns were drawn onto the paper. Water lines were represented by magnets of various sizes. The thermal conductivity was represented by the reciprocal of the resistivity of the paper.

The Acurad system employed a bottom fill system that required a stable flow-front. Logical thought processes and trial and error were used because computerized analysis did not exist yet; however this modeling was the precursor to computerized flow and fill modeling.

The Acurad system was the first die casting process that could successfully cast low-iron aluminum alloys, such as A356 and A357. In a traditional die casting process these alloys would solder to the die. Similarly, Acurad castings could be heat treated and meet the U.S. military specification MIL-A-21180-D.

Finally, the Acurad system employed a patented double shot piston design. The idea was to use a second piston (located within the primary piston) to apply pressure after the shot had partially solidified around the perimeter of the casting cavity and shot sleeve. While the system was not very effective, it did lead the manufacturer of the Acurad machines, Ube Industries, to discover that it was just as effective to apply sufficient pressure at the right time later in the cycle with the primary piston; this is indirect squeeze casting.

Pore-free

When no porosity is allowed in a cast part then the pore-free casting process is used. It is identical to the standard process except oxygen is injected into the die before each shot to purge any air from the mold cavity. This causes small dispersed oxides to form when the molten metal fills the die, which virtually eliminates gas porosity. An added advantage to this is greater strength. Unlike standard die castings, these castings can be heat treated and welded. This process can be performed on aluminium, zinc, and lead alloys.

Vacuum-assisted High-pressure Die Casting

In vacuum assisted high pressure die casting, a.k.a. vacuum high pressure die casting (VHPDC), a vacuum pump removes air and gases from die cavity and metal delivery system before and during injection. Vacuum die casting reduces porosity, allows heat treating and welding, improves surface finish, and can increase strength.

Heated-manifold Direct-injection

Heated-manifold direct-injection die casting, also known as direct-injection die casting or runnerless die casting, is a zinc die casting process where molten zinc is forced through a heated manifold and then through heated mini-nozzles, which lead into the molding cavity. This process has the advantages of lower cost per part, through the reduction of scrap (by the elimination of sprues, gates, and runners) and energy conservation, and better surface quality through slower cooling cycles.

Semi-solid

Semi-solid die casting uses metal that is heated between its liquidus and either solidus or eutectic temperature, so that it is in its "mushy region". This allows for more complex parts and thinner walls.

SHELL MOLDING

Shell moulding, which is sometimes known as shell mold casting, is one of a most suitable casting process for complex and other thin sections. This process gives high accuracy and better surface finish. One of the major advantages of this casting process which make it suitable for various casting industries is that it can easily automated. These all advantages make this process versatile and easily adoptable for any industries.

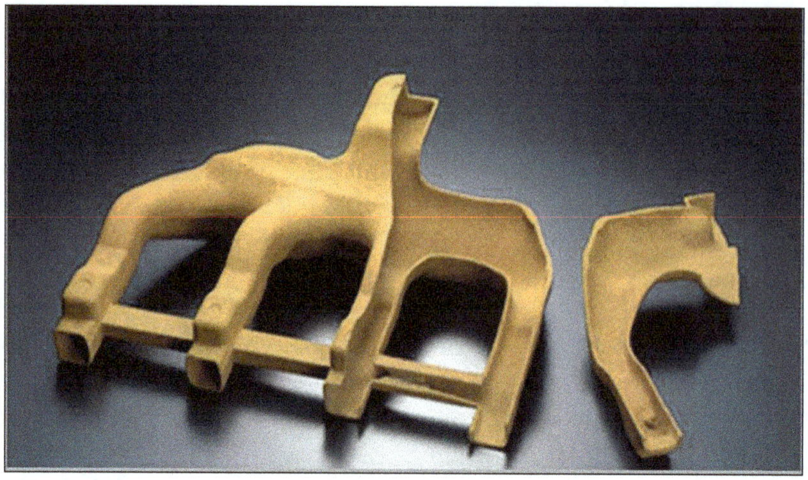

Shell mold casting process uses a mold like a shell, which is made by the mixture of sand and resin. This mold is light weighted and have high strength which makes it suitable for automatic casting. This process is suitable for both ferrous and non-ferrous casting but it is mostly used for steel.

Shell Moulding Process

Shell moulding involves various steps. These are given below:

Making of Metal Pattern

The first and most important step involves in every casting process is pattern making. Pattern is replica of the final product. It can made by wood, plastic, metal etc. Shell moulding uses a metal pattern along with all pattern allowance. This pattern is made by either aluminium of cast iron. One of the major advantage of using metal pattern is that it gives high accurate casting and can be used for multiple casting.

Pattern creation.

Heating of Pattern

The metal pattern created by other casting process is now heated at a temperature range between 180 – 250 degree centigrade. This heating is essential which allows to solidify resin mixed silica when poured on it. After heating, a small layer of lubricant (mosltly Silicon) sprayed over metallic pattern which allows easy removal of metallic pattern from shell.

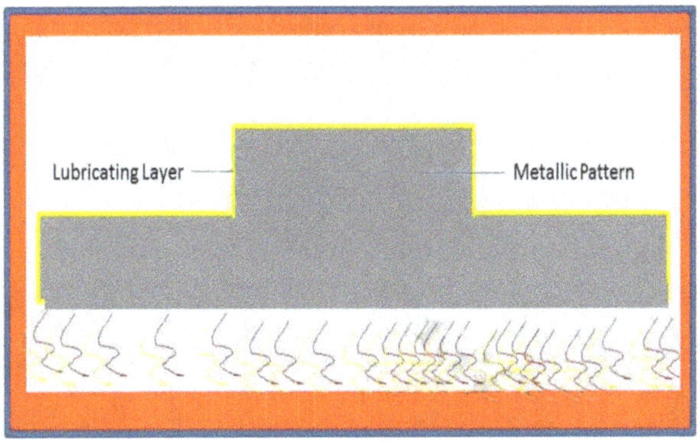

Heating of pattern.

Shell Mould Creation

This step is major step in this casting process. This step can be further divided into following steps:

- Pattern is clamped over a dump box. Now this assembly is turned face down. A mixture of sand and resin is filled into dump box. Mostly fine grade green sand is used for mixture.

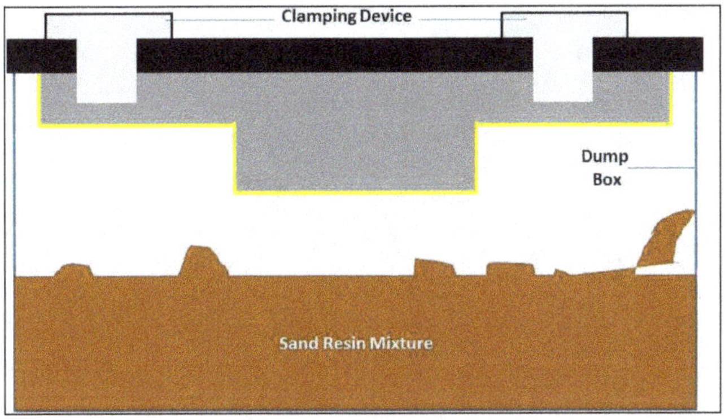

Shell mould creation.

- Now this whole assemble is inverted which allow to sand resin mixture fall over heated pattern. A layer of mixture, which is in direct contact with pattern is become hard and form a shell. The thickness of shell is mostly depends on the temperature of the pattern and time duration of contact.

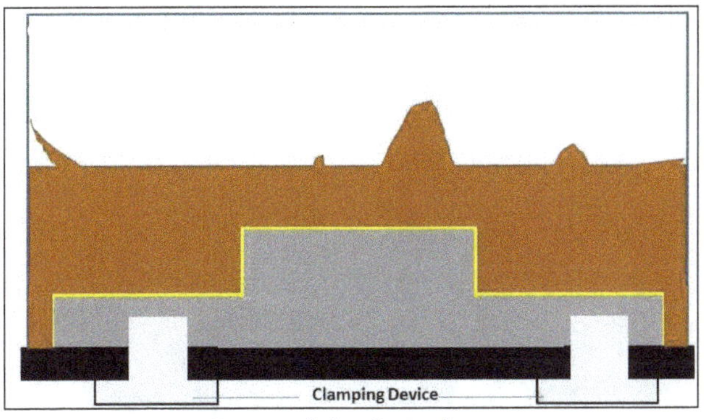

Inverted the dump box.

- Now the dump box is again inverted which allow to remove extra sand resin mixture.

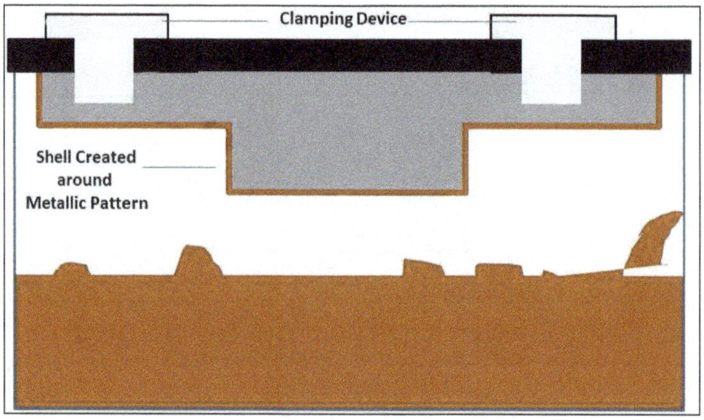

Shell creation.

- After it, metallic pattern is removed from dump box and shell is separated from it. The other half of the mold is also created using same technic.

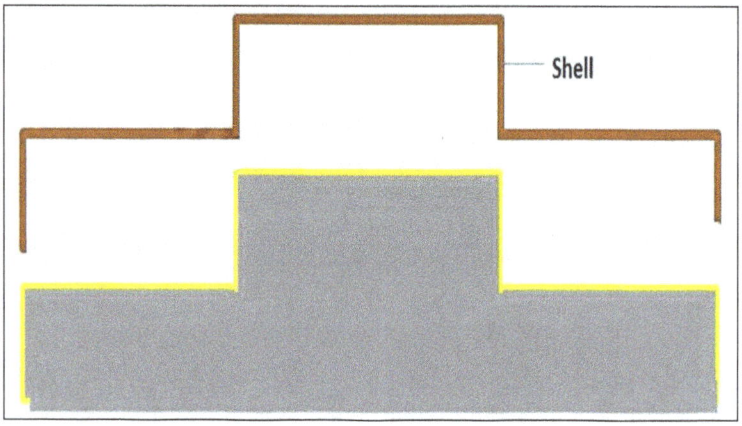

Shell removed metallic pattern.

Mold Assembly

This step assemble all shells created by the shell moulding. The all required shell assembled into a flask and supported by a baking material. All desirable part like cores, runner, riser etc. are attached to it.

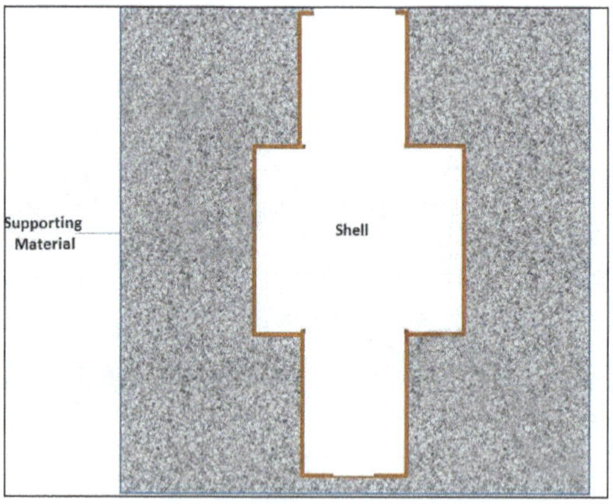

Mould assembled.

Casting Formation

Now the cavity is filled with a molten metal and allow to solidify. After solidification the metal cast is removed by breaking the shell. The casting formed by this process is highly accurate and well finished. Generally it does not required further machining.

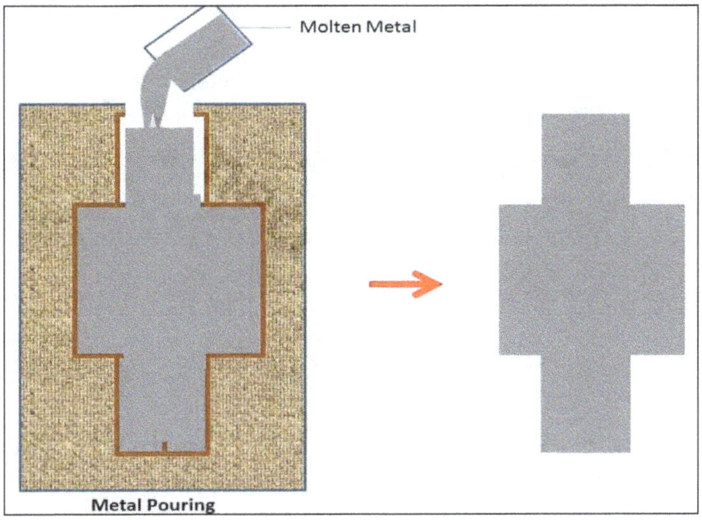

Casting formation.

Advantages and Disadvantages

Advantages

- Thin and complex section can be easily cast.
- High dimension accuracy and good surface finish.
- Easily automated.
- It can be operate by an semi skilled operator.
- No further machining required.

Disadvantages

- Special metal pattern required which makes it expensive for large casting.
- It is not suitable for small batch production.
- Shell moulds are less permeable compare to green sand mould.
- Size and weight limitation.

Application

Most of industrial products like gearbox housing, connecting rod, small size boats, truck hoods, cylindrical head, Camshaft, valve body etc. are made by shell moulding.

References

- Donaldson, Brent (2017-11-01), “Foundry Says Robotic Sand Printing a “Game Changer” for Metal Casting”, Additive Manufacturing, retrieved 2017-11-14
- What-Is-casting, manufacturing-technology: brighthubengineering.com, Retrieved 13 August, 2019
- What-is-centrifugal-casting-and-how-does-it-wor: marketing.metaltek.com, Retrieved 14 January, 2019
- Stephanie Dalley, The Mystery of the Hanging Garden of Babylon: An Elusive World Wonder Traced, Oxford University Press (2013). ISBN 978-0-19-966226-5. Translation by the author, reproduced by permission of Oxford University Press
- Evaporative-pattern-casting-epc-process, advanced-casting-technologies, books: intechopen.com, Retrieved 15 February, 2019
- About die casting, The North American Die Casting Association, archived from the original on 21 October 2010, retrieved 15 October 2010

Molding: Types and Processes

CHAPTER 4

The process of manufacturing which includes the shaping of a liquid or pliable raw material with a rigid frame is known as the molding process. Some of the various types of molding processes include compression molding, blow molding, dip molding, rotational molding, thermoforming, etc. The topics elaborated in this chapter will help in gaining a better perspective about these molding processes.

Molding processes are the major periodic manufacturing operations in the polymer industry. An enormous amount of plastic, rubber, and thermoset parts, ranging from automobile bumpers to personal computers, refrigerator housings to bottles and tires, are produced by molding processes. The automotive, appliance, computer, beverage, tire, and other industries are associated with molding.

Rheology plays a very important role in molding processes. In particular, the shear rate and temperature dependence of viscosity determines the resistance to flow of polymer melts in dies and molds. As flow rate or output of process increases, higher shear rates and accordingly higher shear stresses are developed and higher forces and pressures are required to shape polymer products. The sensitivity of viscosity to variations in shear rate, characterized by the shear thinning effect, determines pressure increase during the mold filling process. This means that polymer melts exhibiting a lower power-law index and, therefore, more shear thinning, and show less increase in pressure with increase of flow rate during processing. Therefore, they will have lower energy consumption. On the other hand, polymer melts exhibiting higher temperature sensitivity of viscosity, i.e., higher activation energy of viscous flow, would cause a faster increase in pressures and forces required to carry out molding processes, which take place under fast cooling rates. Viscosity, and its shear rate and temperature dependence, determine whether a mold is fully filled or not.

Viscoelasticity of polymer melts has a significant influence on polymer molding and the performance of shaped products, especially with respect to a level of frozen-in molecular orientation introduced in them during cavity filling and packing stages and subsequent relaxation processes occurring upon the cooling stage of the molding cycle. In particular, polymer melts exhibiting a higher relaxation time would lead to a higher level of a frozen-in molecular orientation and residual flow stresses in final products. This strongly affects the performance characteristics of final products. Due to these effects, product properties become highly anisotropic. Depending on the type of products made, the molecular orientation effect may be beneficial or detrimental for product performance. For example, in the case of molding optical products, a higher level of molecular orientation introduces higher anisotropy of refractive index, leading to deterioration of their optical quality. In particular, this effect of the frozen-in orientation is well known in the case of injection molding of optical products, such as compact disks, CD, and DVD substrates, and various lenses. A high level of frozen-in birefringence or retardation causes distortion of laser light propagation through CD and DVD media, leading to a poor quality of reproduction of the sound of music and optical pictures. Therefore, very stringent specifications are established concerning the level of residual optical retardation in these molded products. It should also be noted that this optical retardation is strongly affected by process parameters during molding, as well as relaxation characteristics, such as the relaxation times and their distribution, and optical constants of polymer melts, such as the stress-optical and strain-optical coefficients.

Plastics molding is an industry of enormous volume and scope. There are many variations of molding technology. These processes include compression, injection, injection-compression, co-injection, transfer, resin transfer, blow, rotational molding and thermoforming.

COMPRESSION MOLDING

Compression Moulding is a method of moulding in which the moulding material, generally preheated, is first placed in an open, heated mould cavity. The mould is closed with a top force or plug member, pressure is applied to force the material into contact with all mould areas, while heat and pressure are maintained until the moulding material has cured. The process employs thermosetting resins in a partially cured stage, either in the form of granules, putty-like masses, or preforms.

Compression molding is a high-volume, high-pressure method suitable for molding complex, high-strength fiberglass reinforcements. Advanced composite thermoplastics can also be compression molded with unidirectional tapes, woven fabrics, randomly oriented fiber mat or chopped strand. The advantage of compression molding is its ability to mold large, fairly intricate parts. Also, it is one of the lowest cost molding methods compared with other methods such as transfer molding and injection molding; moreover it wastes relatively little material, giving it an advantage when working with expensive compounds.

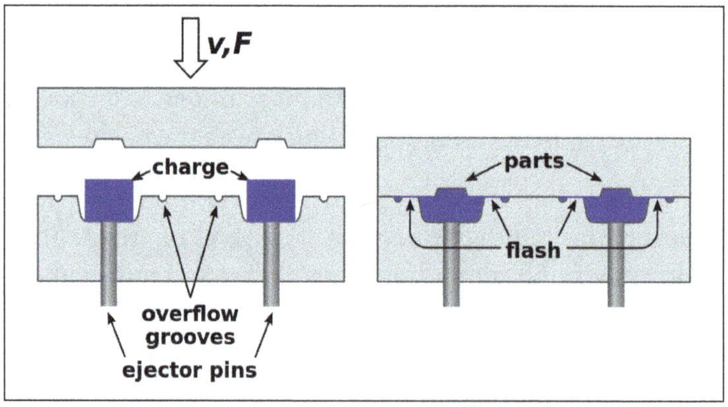

Compression molding - simplified diagram of the process.

However, compression molding often provides poor product consistency and difficulty in controlling flashing, and it is not suitable for some types of parts. Fewer knit lines are produced and a smaller amount of fiber-length degradation is noticeable when compared to injection molding. Compression-molding is also suitable for ultra-large basic shape production in sizes beyond the capacity of extrusion techniques. Materials that are typically manufactured through compression molding include: Polyester fiberglass resin systems (SMC/BMC), Torlon, Vespel, Poly(p-phenylene sulfide) (PPS), and many grades of PEEK.

Compression molded rubber boots before the flashes are removed.

Compression molding is commonly utilized by product development engineers seeking cost effective rubber and silicone parts. Manufacturers of low volume compression molded components include PrintForm, 3D, STYS, and Aero MFG.

Compression molding was first developed to manufacture composite parts for metal replacement applications, compression molding is typically used to make larger flat or moderately curved parts. This method of molding is greatly used in manufacturing automotive parts such as hoods, fenders, scoops, spoilers, as well as smaller more intricate parts. The material to be molded is positioned in the mold cavity and the heated platens are closed by a hydraulic ram. Bulk molding compound (BMC) or sheet molding compound (SMC), are conformed to the mold form by the applied pressure and heated until the curing reaction occurs. SMC feed material usually is cut to conform to the surface area of the mold. The mold is then cooled and the part removed.

Mold and material behavior vary based on material types. New I-PRESS Servo Hydraulic technology provides end users with a greater degree of flexibility with press movement & pressure profiles and control of external devices such as automation to platen zone temperature control. With the latest of electronic servo motor to high pressure hydraulic pump users save considerable energy and the greatest in speed, distance, pressure, dwell time and burping movements.

Servo Hydraulic I-PRESS Compression Molding Controller.

Materials may be loaded into the mold either in the form of pellets or sheet, or the mold may be loaded from a plasticating extruder. Materials are heated above their melting points, formed and cooled. The more evenly the feed material is distributed over the mold surface, the less flow orientation occurs during the compression stage.

Compression molding is also widely used to produce sandwich structures that incorporate a core material such as a honeycomb or polymer foam.

Thermoplastic matrices are commonplace in mass production industries. One significant example are automotive applications where the leading technologies are long fibre reinforced thermoplastics (LFT) and glass fiber mat reinforced thermoplastics (GMT).

In compression molding there are six important considerations that an engineer should bear in mind:

- Determining the proper amount of material.
- Determining the minimum amount of energy required to heat the material.
- Determining the minimum time required to heat the material.
- Determining the appropriate heating technique.
- Predicting the required force, to ensure that shot attains the proper shape.
- Designing the mold for rapid cooling after the material has been compressed into the mold.

Compression molding is a forming process in which a plastic material is placed directly into a heated metal mold then is softened by the heat and therefore forced to conform to the shape of the mold, as the mold closes. Once molding is completed excess flash may be removed. Typically, compression molding machines open along a vertical axis.

Process Characteristics

The use of thermoset plastic compounds characterizes this molding process from many of the other molding processes. These thermosets can be in either preform or granule shapes. Unlike some of the other processes we find that the materials are usually preheated and measured before molding. This helps to reduce excess flash. Inserts, usually metallic, can also be molded with the plastic. As a side note, remember not to allow any undercuts on the shape, it will make ejection especially difficult. Thermoplastic matrices with an inherent indefinite shelf-life and shorter cycle moulding times are widely used.

Process Schematic

Compression molding is one of the oldest manufacturing technique for rubber molding. The process parameters includes molding time, temperature, and pressure. Usually, a 300-400 ton clamp pressure is used. The typical mold is shaped like a clam shell. The molding press looked a lot like a ladle filled vertical press used for casting aluminum. The bottom of the mold was always the cavity. Compression molding used preforms made by an extruder/wink cutter or a roller die/die cutter. Wink meaning that 2 blades meet on center to cut the extrudate to length. For example;

Molding water bottles used die cut sheets from a roller die. The sheet was 3 inch by 6 inch. The first sheet was placed- one below a core and one sheet of equal size above the core, and then the top of the mold lowered by hand or by hoist to near shut. The mold was then pushed into the press. The start button hydraulically closed the vertical press to full pressure. The mold temperature was about 350 degrees. The platens of the presses were steam heated. When the cycle ended (about 3.5-4.0 Minutes) the press would open and the mold would be pulled out toward the operator. The operator would try to open the clam shell mold top, and then lean the top of the mold back against the press. Exposed is the bottle with the core still inside. While the bottle was still hot the operator would insert prongs like reverse pliers in between the bottle rubber and the steel core. The operator would then stretch the bottle at the neck over the core to free the bottle. In preparation of compression molding baby nipples and golf ball centers the preforms were extruded. The baby nipple was a kidney shape about 2 inches tall and 1/2 inch wide in the middle. The golf ball center preform had a 1 x 1 inch round slug. Both slugs were designed to stand up in the mold cavity. During the cycle the operator would load the jig with slugs. When the mold is opened, the lower platen would lower and the mold would be hydraulically pushed ou to the operator. Therein, the heat sheet (all molded parts from that cycle were joined together by a parting line rind (flash)) were placed in a transfer cart. The next cycle began by the jig being put over the mold. The slide tray was pulled and the preforms were released into the cavity of the mold. The start button moved the lower platen back into the press and the cure cycle bagan again. Therein the first cycle was complete. Each operator ran an average of 4 presses. Loading an unload was done during the cycle. The heat sheets removed from the mold were then transported to a die station. The die out station would remove the rind leaving the finished parts. With the evolution of compression molding, next was injection transfer. Basically the extruder was made part of the molding cycle. The rubber was injected into an upper heated platen station, and then pressure was applied to transfer molten rubber to the clamped mold. The design of injection transfer and improved molds were more so plastic injection molding except the platens and molds of injection transfer are heated. In contrast plastic Injections molding shoots a hot plastic into a cold mold.

Typical Tools and Geometry Produced

Three types of molds used are the flash plunger-type, straight plunger-type, and the "landed" plunger-type molds. The flash type mold must have an accurate charge of plastic and produces a horizontal flash (excess material protruding from the mold). The straight plunger-type mold allows for some inaccuracy in the charge of plastic and produces a vertical flash. The landed plunger type mold must have an accurate charge of plastic, and no flash is produced.

BLOW MOLDING

Blow molding (BrE moulding) is a specific manufacturing process by which hollow plastic parts are formed and can be joined together. It is also used for forming glass bottles or other hollow shapes.

In general, there are three main types of blow molding extrusion blow molding, injection blow molding, and injection stretch blow molding.

The blow molding process begins with melting down the plastic and forming it into a parison or, in the case of injection and injection stretch blow molding (ISB), a preform. The parison is a tube-like piece of plastic with a hole in one end through which compressed air can pass.

The parison is then clamped into a mold and air is blown into it. The air pressure then pushes the plastic out to match the mold. Once the plastic has cooled and hardened the mold opens up and the part is ejected. The cost of blow molded parts is higher than that of injection-molded parts but lower than rotational molded parts.

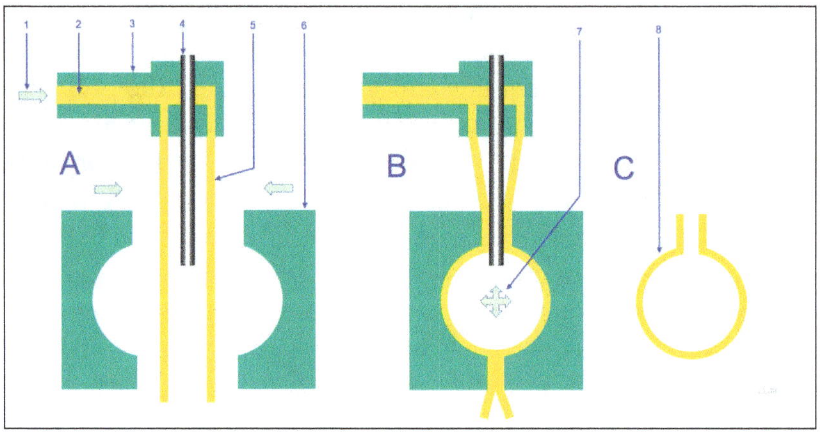

The blow molding process.

Typologies

Extrusion Blow Molding

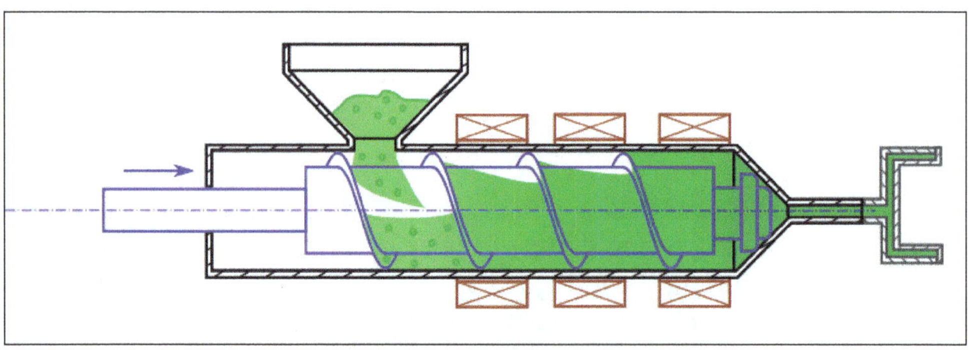
Extrusion blow molding.

In extrusion blow molding (EBM), plastic is melted and extruded into a hollow tube (a parison). This parison is then captured by closing it into a cooled metal mold. Air is then blown into the parison, inflating it into the shape of the hollow bottle, container, or part. After the plastic has cooled sufficiently, the mold is opened and the part is ejected. Continuous and Intermittent are two variations of Extrusion Blow Molding. In continuous extrusion blow molding the parison is extruded continuously and the individual parts are cut off by a suitable knife. In Intermittent blow molding there are two processes: straight intermittent is similar to injection molding whereby the screw turns, then stops and pushes the melt out. With the accumulator method, an accumulator gathers melted plastic and when the previous mold has cooled and enough plastic has accumulated, a rod pushes the melted plastic and forms the parison. In this case the screw may turn continuously or intermittently. With continuous extrusion the weight of the parison drags the parison and makes calibrating the wall thickness difficult. The accumulator head or reciprocating screw methods use hydraulic systems to push the parison out quickly reducing the effect of the weight and allowing precise control over the wall thickness by adjusting the die gap with a parison programming device.

EBM processes may be either continuous (constant extrusion of the parison) or intermittent. Types of EBM equipment may be categorized as follows:

Continuous Extrusion Equipment

- Rotary wheel blow molding systems.
- Shuttle machinery.

Intermittent Extrusion Machinery

- Reciprocating screw machinery.
- Accumulator head machinery.

Examples of parts made by the EBM process include most polyethylene hollow products, milk bottles, shampoo bottles, automotive ducting, watering cans and hollow industrial parts such as drums.

Advantages of blow molding include: low tool and die cost; fast production rates; ability to mold complex part; Handles can be incorporated in the design.

Disadvantages of blow molding include: limited to hollow parts, low strength, to increase barrier properties multilayer parisons of different materials are used thus not recyclable. To make wide neck jars spin trimming is necessary.

Spin Trimming

Containers such as jars often have an excess of material due to the molding process. This is trimmed off by spinning a knife around the container which cuts the material away. This excess plastic is then recycled to create new moldings. Spin Trimmers are used on a number of materials, such as PVC, HDPE and PE+LDPE. Different types of the materials have their own physical characteristics affecting trimming. For example, moldings produced from amorphous materials are much more difficult to trim than crystalline materials. Titanium coated blades are often used rather than standard steel to increase life by a factor of 30 times.

Injection Blow Molding

The process of injection blow molding (IBM) is used for the production of hollow glass and plastic objects in large quantities. In the IBM process, the polymer is injection molded onto a core pin; then the core pin is rotated to a blow molding station to be inflated and cooled. This is the least-used of the three blow molding processes, and is typically used to make small medical and single serve bottles. The process is divided into three steps: injection, blowing and ejection.

The injection blow molding machine is based on an extruder barrel and screw assembly which melts the polymer. The molten polymer is fed into a hot runner manifold where it is injected through nozzles into a heated cavity and core pin. The cavity mold forms the external shape and is clamped around a core rod which forms the internal shape of the preform. The preform consists of a fully formed bottle/jar neck with a thick tube of polymer attached, which will form the body, similar in appearance to a test tube with a threaded neck.

The preform mold opens and the core rod is rotated and clamped into the hollow, chilled blow mold. The end of the core rod opens and allows compressed air into the preform, which inflates it to the finished article shape.

After a cooling period the blow mold opens and the core rod is rotated to the ejection position. The finished article is stripped off the core rod and as an option can be leak-tested prior to packing. The preform and blow mold can have many cavities, typically three to sixteen depending on the article size and the required output. There are three sets of core rods, which allow concurrent preform injection, blow molding and ejection.

Advantages: It produces an injection molded neck for accuracy.

Disadvantages: Only suits small capacity bottles as it is difficult to control the base centre during blowing. No increase in barrier strength as the material is not biaxially stretched. Handles can't be incorporated.

Injection Stretch Blow Molding Process

This has two main different methods, namely Single-stage and two-stage process. Single-stage process is again broken down into 3-station and 4-station machines. In the two-stage injection stretch blow molding process, the plastic is first molded into a "preform" using the injection molding process. These preforms are produced with the necks of the bottles, including threads (the "finish") on one end. These preforms are packaged, and fed later (after cooling) into a re-heat stretch blow molding machine. In the ISB process, the preforms are heated (typically using infrared heaters) above their glass transition temperature, then blown using high-pressure air into bottles using metal blow molds. The preform is always stretched with a core rod as part of the process.

Advantages: Very high volumes are produced. Little restriction on bottle design. Preforms can be sold as a completed item for a third party to blow. Is suitable for cylindrical, rectangular or oval bottles.

Disadvantages: High capital cost. Floor space required is high, although compact systems have become available.

In the single-stage process both preform manufacture and bottle blowing are performed in the same machine. The older 4-station method of injection, reheat, stretch blow and ejection is more costly than the 3-station machine which eliminates the reheat stage and uses latent heat in the pre-form, thus saving costs of energy to reheat and 25% reduction in tooling. The process explained: Imagine the molecules are small round balls, when together they have large air gaps and small surface contact, by first stretching the molecules vertically then blowing to stretch horizontally the biaxial stretching makes the molecules a cross shape. These "crosses" fit together leaving little space as more surface area is contacted thus making the material less porous and increasing barrier strength against permeation. This process also increases the strength to be ideal for filling with carbonated drinks.

Advantages

Highly suitable for low volumes and short runs. As the preform is not released during the entire process the preform wall thickness can be shaped to allow even wall thickness when blowing rectangular and non-round shapes.

Disadvantages

Restrictions on bottle design. Only a champagne base can be made for carbonated bottles.

DIP MOLDING

Dip molding is a process used to create a layer of plastic, usually polyvinyl chloride (PVC), over a metal object. Whether it's steel, iron, copper, brass or aluminum, metal has different properties than plastic. Unlike plastic, for example, must metal corrodes when exposed to moisture. To protect against certain metal objects from corrosion — while also improving other characteristics of metal objects — manufacturers use dip molding.

Steps for Dip Molding

To dip mold an object, the mold is first heated and then submerged into the PVC material. Next, the mold is removed, drained and left to dry. Depending on the specific type of dip molding process used, the mold may then be cured. After the mold has finished drying and/or curing, the PVC shell be removed.

It's important to note that dip molding often requires several cycles of submersion in the PCV material. If the PVC shell is too thin for the desired application, for example, the mold may be re-submerged into the PVC material. Each submersion adds a little more PVC material to the mold, making it thicker and thereby stronger and more durable.

Advantages and Disadvantages of Dip Molding

Dip molding has both advantages and disadvantages over other molding processes. It's relatively inexpensive, for instance, allowing manufacturing companies to add a protective layer of plastic over their products or components without spending a fortune. The finished PVC shell is also easy to remove from the mold, as PVC material is naturally soft and somewhat elastic.

On the other hand, dip molding can be a time-consuming and tedious molding process. The PVC material must first be heated, and after dipping the mold into it, the mold be left to dry and/or cure. Furthermore, some manufacturing companies may have to submerge the mold into the liquid PVC several times to achieve the desired thickness.

ROTATIONAL MOLDING

Rotational Molding (BrE moulding) involves a heated hollow mold which is filled with a charge or shot weight of material. It is then slowly rotated (usually around two perpendicular axes), causing the softened material to disperse and stick to the walls of the mold. In order to maintain even thickness throughout the part, the mold continues to rotate at all times during the heating phase and to avoid sagging or deformation also during the cooling phase. The process was applied to plastics in the 1950s but in the early years was little used because it was a slow process restricted to a small number of plastics. Over time, improvements in process control and developments with plastic powders have resulted in a significant increase in usage.

Rotocasting (also known as rotacasting), by comparison, uses self-curing resins in an unheated mould, but shares slow rotational speeds in common with rotational molding. Spin casting should not be confused with either, utilizing self-curing resins or white metal in a high-speed centrifugal casting machine.

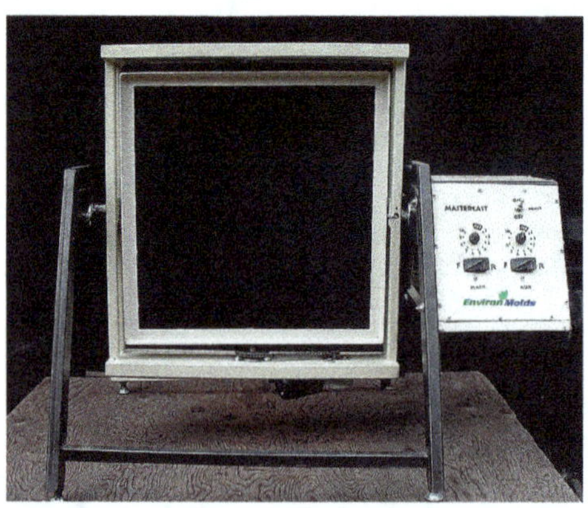

A three-motor powered (tri-power) rotational-molding or spin-casting machine.

Equipment and Tooling

Rotational molding machines are made in a wide range of sizes. They normally consist of molds, an oven, a cooling chamber, and mold spindles. The spindles are mounted on a rotating axis, which provides a uniform coating of the plastic inside each mold.

Molds (or tooling) are either fabricated from welded sheet steel or cast. The fabrication method is often driven by part size and complexity; most intricate parts are likely made out of cast tooling. Molds are typically manufactured from stainless steel or aluminum. Aluminum molds are usually much thicker than an equivalent steel mold, as it is a softer metal. This thickness does not affect cycle times significantly since aluminum's thermal conductivity is many times greater than steel. Due to the need to develop a model prior to casting, cast molds tend to have additional costs associated with the manufacturing of the tooling, whereas fabricated steel or aluminum molds, particularly when used for less complex parts, are less expensive. However, some molds contain both aluminum and steel. This allows for variable thicknesses in the walls of the product. While this process is not as precise as injection molding, it does provide the designer with more options. The aluminum addition to the steel provides more heat capacity, causing the melt-flow to stay in a fluid state for a longer period.

Standard Setup and Equipment for Rotational Molding

Normally all rotation molding systems have a number of parts including molds, oven, cooling chamber and mold spindles. The molds are used to create the part, and are typically made of aluminium. The quality and finish of the product is directly related to the quality of the mold being used. The oven is used to heat the part while also rotating the part to form the part desired. The cooling chamber is where the part is placed until it cools, and the spindles are mounted to rotate and provide a uniform coat of plastic inside each mold.

Rotational Molding Machines

Rock and Roll Machine

A Rock and Roll rotational molding machine built.

This is a specialized machine designed mainly to produce long narrow parts. Some are of the clamshell type, thus one arm, but there are also shuttle-type Rock & Roll machines, with two arms. Each arm rotates or rolls the mold 360 degrees in one direction and at the same time tips and rocks the mold 45 degrees above or below horizontal in the other direction. Newer machines use forced hot air to heat the mold. These machines are best for large parts that have a large length-to-width ratio. Because of the smaller heating chambers, there is a saving in heating costs compared to bi-axial machines.

Clamshell Machine

This is a single arm rotational molding machine. The arm is usually supported by other arms on both ends. The clamshell machine heats and cools the mold in the same chamber. It takes up less space than equivalent shuttle and swing arm rotational molders. It is low in cost compared to the size of products made. It is available in smaller scales for schools interested in prototyping and for high quality models. More than one mold can be attached to the single arm.

Vertical or Up and Over Rotational Machine

The loading and unloading area is at the front of the machine between the heating and cooling areas. These machines vary in size between small to medium compared to other rotational machines. Vertical rotational molding machines are energy efficient due to their compact heating and cooling chambers. These machines have the same (or similar) capabilities as the horizontal carousel multi-arm machines, but take up much less space.

Shuttle Machine

Most shuttle machines have two arms that move the molds back and forth between the heating chamber and cooling station. The arms are independent of each other and they turn the molds bi-axially. In some cases, the shuttle machine has only one arm. This machine moves the mold in a linear direction in and out of heating and cooling chambers. It is low in cost for the size of product produced and the footprint is kept to a minimum compared to other types of machines. It is also available in smaller scale for schools and prototyping.

Swing Arm Machine

The swing-arm machine can have up to four arms, with a bi-axial movement. Each arm is independent from each other as it is not necessary to operate all arms at the same time. Each arm is mounted on a corner of the oven and it swings in and out of the oven. On some swing-arm machines, a pair of arms is mounted on the same corner, thus a four-arm machine has two pivot points. These machines are very useful for companies that have long cooling cycles or require a lot of time to demold parts, compared to the cook time. It is a lot easier to schedule maintenance work or try to run a new mold without interrupting production on the other arms of the machine.

Carousel Machine

A Carousel machine with four independent arms.

This is one of the most common bi-axial machines in the industry. It can have up to 4 arms and six stations and it comes in a wide range of sizes. The machine comes in two different models, fixed and independent. A fixed-arm carousel consists of 3 fixed arms that must move together. One arm will be in the heating chamber while the other is in the cooling chamber and the other in the loading/ reloading area. The fixed-arm carousel works well when working with identical cycle times on each arm. The independent-arm carousel machine is available with 3 or 4 arms that can move separately from the other. This allows for different size molds, with different cycle times and thickness needs.

Production Process

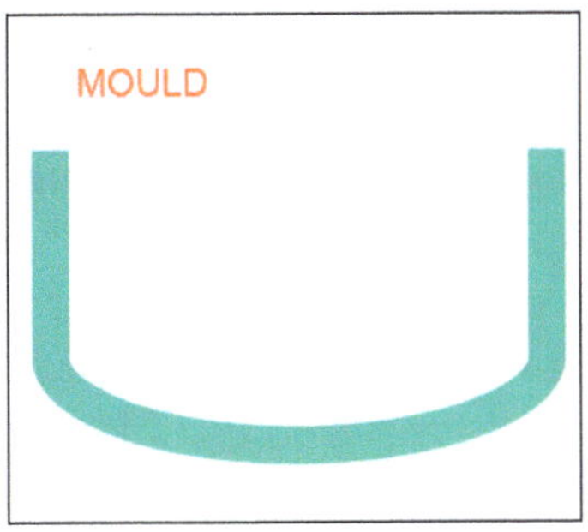

Unloading a molded polyethylene tank in a Shuttle machine.

The rotational molding process is a high-temperature, low-pressure plastic-forming process that uses heat and biaxial rotation (i.e., angular rotation on two axes) to produce hollow, one-piece parts. Critics of the process point to its long cycle times—only one or two cycles an hour can typically occur, as opposed to other processes such as injection molding, where parts can be made in a few seconds. The process does have distinct advantages. Manufacturing large, hollow parts such as oil tanks is much easier by rotational molding than any other method. Rotational molds are significantly cheaper than other types of mold. Very little material is wasted using this process, and excess material can often be re-used, making it a very economically and environmentally viable manufacturing process.

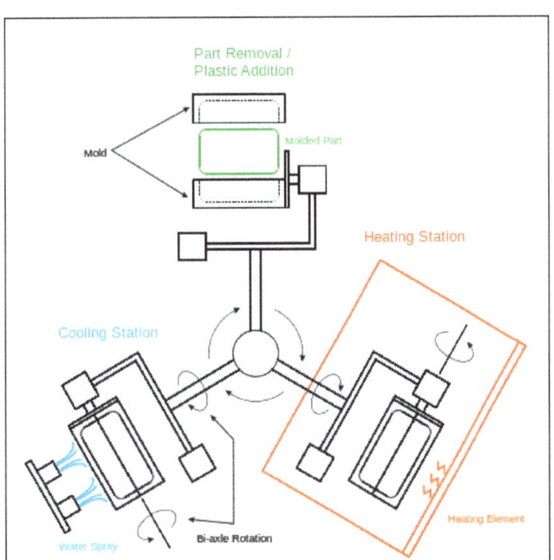

Rotational Molding Process.

The rotational molding process consists of four distinct phases:

- Loading a measured quantity of polymer (usually in powder form) into the mold.
- Heating the mold in an oven while it rotates, until all the polymer has melted and adhered to the mold wall. The hollow part should be rotated through two or more axes, rotating at different speeds, in order to avoid the accumulation of polymer powder. The length of time the mold spends in the oven is critical: too long and the polymer will degrade, reducing impact strength. If the mold spends too little time in the oven, the polymer melt may be incomplete. The polymer grains will not have time to fully melt and coalesce on the mold wall, resulting in large bubbles in the polymer. This has an adverse effect on the mechanical properties of the finished product.
- Cooling the mold, usually by fan. This stage of the cycle can be quite lengthy. The polymer must be cooled so that it solidifies and can be handled safely by the operator. This typically takes tens of minutes. The part will shrink on cooling, coming away from the mold, and facilitating easy removal of the part. The cooling rate must be kept within a certain range. Very rapid cooling (for example, water spray) would result in cooling and shrinking at an uncontrolled rate, producing a warped part.
- Removal of the part.

Until recently, the process largely relied on both trial and error and the experience of the operator to determine when the part should be removed from the oven and when it was cool enough to be removed from the mold. Technology has improved in recent years, allowing the air temperature in the mold to be monitored, removing much of the guesswork from the process.

Much of the current research is into reducing the cycle time, as well as improving part quality. The most promising area is in mold pressurization. It is well known that applying a small amount of pressure internally to the mold at the correct point in the heating phase accelerates coalescence of the polymer particles during the melting, producing a part with fewer bubbles in less time than at atmospheric pressure. This pressure delays the separation of the part from the mold wall due to shrinkage during the cooling phase, aiding cooling of the part. The main drawback to this is the danger to the operator of explosion of a pressurized part. This has prevented adoption of mold pressurization on a large scale by rotomolding manufacturers.

Mold Release Agents

A good mold release agent (MRA) will allow the material to be removed quickly and effectively. Mold releases can reduce cycle times, defects, and browning of finished product. There are a number of mold release types available; they can be categorized as follows:

- Sacrificial coatings: The coating of MRA has to be applied each time because most of the MRA comes off on the molded part when it releases from the tool. Silicones are typical MRA compounds in this category.
- Semi-permanent coatings: The coating, if applied correctly, will last for a number of releases before requiring to be re-applied or touched up. This type of coating is most prevalent

in today's rotational molding industry. The active chemistry involved in these coatings is typically a polysiloxane.

- Permanent coatings: Most often some form of PTFE coating, which is applied to the mold. Permanent coatings avoid the need for operator application, but may become damaged by misuse.

Materials

More than 80% of all the material used is from the polyethylene family: cross-linked polyethylene (PEX), low-density polyethylene (LDPE), linear low-density polyethylene (LLDPE), high-density polyethylene (HDPE), and regrind. Other compounds are PVC plastisols, nylons, and polypropylene.

Order of materials most commonly used by industry:

- Polyethylene,
- Polypropylene,
- Polyvinyl chloride,
- Nylon,
- Polycarbonate.

These materials are also occasionally used (not in order of most used):

- Aluminum,
- Acrylonitrile butadiene styrene (ABS),
- Acetal,
- Acrylic,
- Epoxy,
- Fluorocarbons,
- Ionomer,
- Polybutylene,
- Polyester,
- Polystyrene,
- Polyurethane,
- Silicone,
- Various foods (especially chocolate).

Natural Materials

Recently it has become possible to use natural materials in the molding process. Through the use of real sands and stone chip, sandstone composite can be created which is 80% natural non-processed material.

Rotational molding of plaster is used to produce hollow statuettes.

Chocolate is rotationally molded to form hollow treats.

Products

Designers can select the best material for their application, including materials that meet U.S. Food and Drug Administration (FDA) requirements. Additives for weather resistance, flame retardation, or static elimination can be incorporated. Inserts, graphics, threads, handles, minor undercuts, flat surfaces without draft angles, or fine surface detail can be part of the design. Designs can also be multi-wall, either hollow or foam filled.

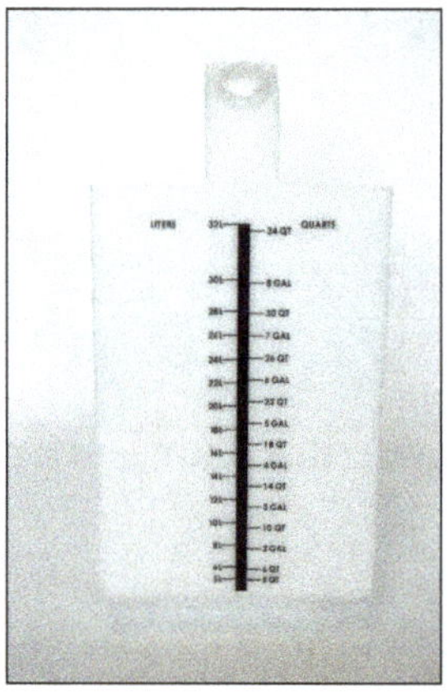

Mold in graphic.

Products that can be manufactured using rotational molding include storage tanks, furniture, road signs and bollards, planters, pet houses, toys, bins and refuse containers, doll parts, road cones, footballs, helmets, canoes, rowing boats, tornado shelters, kayak hulls, underground cellars for vine and vegetables storage and playground slides. The process is also used to make highly specialised products, including UN-approved containers for the transportation of nuclear fissile materials, anti-piracy ship protectors, seals for inflatable oxygen masks and lightweight components for the aerospace industry.

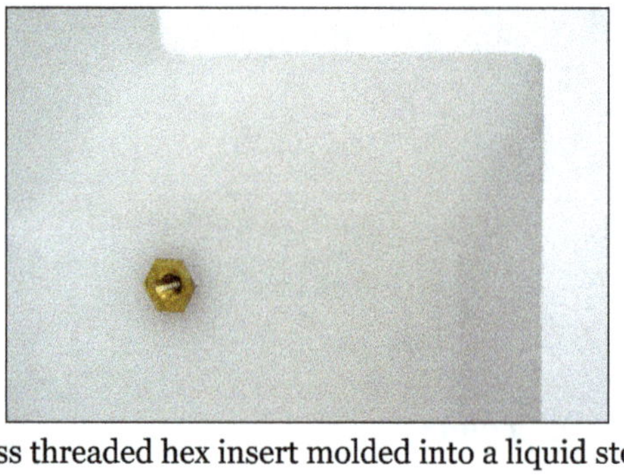

A blind brass threaded hex insert molded into a liquid storage tank.

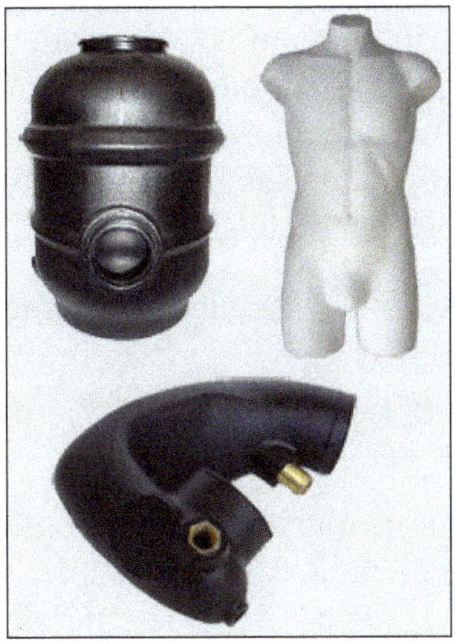

Rotational Molded Flamingo.

Design Considerations

Designing for Rotational Molding

Another consideration is in the draft angles. These are required to remove the piece from the mold. On the outside walls, a draft angle of 1° may work (assuming no rough surface or holes). On inside walls, such as the inside of a boat hull, a draft angle of 5° may be required. This is due to shrinkage and possible part warping.

Another consideration is of structural support ribs. While solid ribs may be desirable and achievable in injection molding and other processes, a hollow rib is the best solution in rotational molding. A solid rib may be achieved through inserting a finished piece in the mold but this adds cost.

Rotational molding excels at producing hollow parts. However, care must be taken when this is done. When the depth of the recess is greater than the width there may be problems with even heating and cooling. Additionally, enough room must be left between the parallel walls to allow for the melt-flow to move properly throughout the mold. Otherwise webbing may occur. A desirable parallel wall scenario would have a gap at least three times the nominal wall thickness, with five times the nominal wall thickness being optimal. Sharp corners for parallel walls must also be considered. With angles of less than 45° bridging, webbing, and voids may occur.

Material Limitations and Considerations

Another consideration is the melt-flow of materials. Certain materials, such as nylon, will require larger radii than other materials. Additionally, the stiffness of the set material may be a factor. More structural and strengthening measures may be required when a flimsy material is used.

Wall Thickness

One benefit of rotational molding is the ability to experiment, particularly with wall thicknesses. Cost is entirely dependent on wall thickness, with thicker walls being costlier and more time consuming to produce. While the wall can have nearly any thickness, designers must remember that the thicker the wall, the more material and time will be required, increasing costs. In some cases, the plastics may significantly degrade due to extended periods at high temperature. Also, different materials have different thermal conductivity, meaning they require different times in the heating chamber and cooling chamber. Ideally, the part will be tested to use the minimum thickness required for the application. This minimum will then be established as a nominal thickness.

For the designer, while variable thicknesses are possible, a process called stop rotation is required. This process is limited in that only one side of the mold may be thicker than the others. After the mold is rotated and all the surfaces are sufficiently coated with the melt-flow, the rotation stops and the melt-flow is allowed to pool at the bottom of the mold cavity.

Wall thickness is important for corner radii as well. Large outside radii are preferable to small radii. Large inside radii are also preferable to small inside radii. This allows for a more even flow of material and a more even wall thickness. However, an outside corner is generally stronger than an inside corner.

Process: Advantages, Limitations and Material Requirements

Advantages

Rotational molding offers design advantages over other molding processes. With proper design, parts assembled from several pieces can be molded as one part, eliminating high fabrication costs. The process also has inherent design strengths, such as consistent wall thickness and strong outside corners that are virtually stress free. For additional strength, reinforcing ribs can be designed into the part. Along with being designed into the part, they can be added to the mold.

The ability to add prefinished pieces to the mold alone is a large advantage. Metal threads, internal pipes and structures, and even different colored plastics can all be added to the mold prior to the addition of plastic pellets. However, care must be taken to ensure that minimal shrinkage while cooling will not damage the part. This shrinking allows for mild undercuts and negates the need for ejection mechanisms (in most pieces).

In some cases rotational molding can be used as a feasible alternative to blow molding, this is due to the similarity in product outputs, with products such as plastic bottles and cylindrical containers, this is only effective on a smaller scale as it much more costly to blow mold regarding a small output, and with fewer resulting products rotational molding is much cheaper, due to blow molding relying on economies of scale regarding efficiency.

Another advantage lies in the molds themselves. Since they require less tooling, they can be manufactured and put into production much more quickly than other molding processes. This is especially true for complex parts, which may require large amounts of tooling for other molding processes. Rotational molding is also the process of choice for short runs and rush deliveries. The molds can be swapped quickly or different colors can be used without purging the mold. With other processes, purging may be required to swap colors.

Due to the uniform thicknesses achieved, large stretched sections are nonexistent, which makes large thin panels possible (although warping may occur). Also, there is little flow of plastic (stretching) but rather a placing of the material within the part. These thin walls also limit cost and production time.

Another cost advantage with rotational molding is the minimal amount of material wasted in production. There are no sprues or runners (as in injection molding), no off-cuts, or pinch off scrap (blow molding). What material is wasted, through scrap or failed part testing, can usually be recycled.

Limitations

Rotationally molded parts have to follow some restrictions that are different from other plastic processes. As it is a low pressure process, sometimes designers face hard to reach areas in the mold. Good quality powder may help overcome some situations, but usually the designers have to keep in mind that it is not possible to make sharp threads that would be possible with injection molding. Some products based on polyethylene can be put in the mold before filling it with the main material. This can help to avoid holes that otherwise would appear in some areas. This could also be achieved using molds with movable sections.

Another limitation lies in the molds themselves. Unlike other processes where only the product needs to be cooled before being removed, with rotational molding the entire mold must be cooled. While water cooling processes are possible, there is still a significant down time of the mold. Additionally, this increases both financial and environmental costs. Some plastics will degrade with the long heating cycles or in the process of turning them into a powder to be melted.

The stages of heating and cooling involve transfer of heat first from the hot medium to the polymer material and next from it to the cooling environment. In both cases, the process of heat transfer occurs in an unsteady regime; therefore, its kinetics attracts the greatest interest in considering these steps. In the heating stage, the heat taken from the hot gas is absorbed both by the mold and the polymer material. The rig for rotational molding usually has a relatively small wall thickness and is manufactured from metals with a high thermal conductivity (aluminum, steel). As a rule, the mold transfers much more heat than plastic can absorb; therefore, the mold temperature must vary linearly. The rotational velocity in rotational molding is rather low (4 to 20 rpm). As a result, in the first stages of the heating cycle, the charged material remains as a powder layer at the bottom of the mold. The most convenient way of changing the cycle is by applying PU sheets in hot rolled forms.

Material Requirements

Due to the nature of the process, materials selection must take into account the following:

- Due to high temperatures within the mold the plastic must have a high resistance to permanent change in properties caused by heat (high thermal stability).
- The molten plastic will come into contact with the oxygen inside the mold—this can potentially lead to oxidation of the melted plastic and deterioration of the material's properties. Therefore, the chosen plastic must have a sufficient amount of antioxidant molecules to prevent such degradation in its liquid state.

- Because there is no pressure to push the plastic into the mold, the chosen plastic must be able to flow easily through the cavities of the mold. The part's design must also take into account the flow characteristics of the particular plastic chosen.

Rotational moulding is basically used for big storage tanks.

THERMOFORMING

Thermoforming is a manufacturing process where a plastic sheet is heated to a pliable forming temperature, formed to a specific shape in a mold, and trimmed to create a usable product. The sheet, or "film" when referring to thinner gauges and certain material types, is heated in an oven to a high-enough temperature that permits it to be stretched into or onto a mold and cooled to a finished shape. Its simplified version is vacuum forming.

In its simplest form, a small tabletop or lab size machine can be used to heat small cut sections of plastic sheet and stretch it over a mold using vacuum. This method is often used for sample and prototype parts. In complex and high-volume applications, very large production machines are utilized to heat and form the plastic sheet and trim the formed parts from the sheet in a continuous high-speed process and can produce many thousands of finished parts per hour depending on the machine and mold size and the size of the parts being formed.

Thermoforming differs from injection molding, blow molding, rotational molding and other forms of processing plastics. Thin-gauge thermoforming is primarily the manufacture of disposable cups, containers, lids, trays, blisters, clamshells, and other products for the food, medical, and general retail industries. Thick-gauge thermoforming includes parts as diverse as vehicle door and dash panels, refrigerator liners, utility vehicle beds and plastic pallets.

In the most common method of high-volume, continuous thermoforming of thin-gauge products, plastic sheet is fed from a roll or from an extruder into a set of indexing chains that incorporate pins, or spikes, that pierce the sheet and transport it through an oven for heating to forming temperature. The heated sheet then indexes into a form station where a mating mold and pressure-box close on the sheet, with vacuum then applied to remove trapped air and to pull the material into or onto the mold along with pressurized air to form the plastic to the detailed shape of the mold. (Plug-assists are typically used in addition to vacuum in the case of taller, deeper-draw formed parts in order to provide the needed material distribution and thicknesses in the finished parts.) After a short form cycle, a burst of reverse air pressure is actuated from the vacuum side of the mold as the form tooling opens, commonly referred to as air-eject, to break the vacuum and assist the formed parts off of, or out of, the mold. A stripper plate may also be utilized on the mold as it opens for ejection of more detailed parts or those with negative-draft, undercut areas. The sheet containing the formed parts then indexes into a trim station on the same machine, where a die cuts the parts from the remaining sheet web or indexes into a separate trim press where the formed parts are trimmed. The sheet web remaining after the formed parts are trimmed is typically wound onto a take-up reel or fed into an inline granulator for recycling.

Most thermoforming companies recycle their scrap and waste plastic, either by compressing in a baling machine or by feeding into a granulator (grinder) and producing ground flake, for sale to reprocessing companies or re-use in their own facility. Frequently, scrap and waste plastic from the thermoforming process is converted back into extruded sheet for forming again.

Thin-gauge and Heavy-gauge (Thick) Thermoforming

There are two general thermoforming process categories. Sheet thickness less than 1.5 mm (0.060 inches) is usually delivered to the thermoforming machine from rolls or from a sheet extruder. Thin-gauge roll-fed or inline extruded thermoforming applications are dominated by rigid or semi-rigid disposable packaging. Sheet thicknesses greater than 3 mm (0.120 inches) are usually delivered to the forming machine by hand or an auto-feed method already cut to final dimensions. Heavy, or thick-gauge, cut sheet thermoforming applications are primarily used as permanent structural components. There is a small but growing medium-gauge market that forms sheet 1.5 mm to 3 mm in thickness.

Heavy-gauge forming utilizes the same basic process as continuous thin-gauge sheet forming, typically draping the heated plastic sheet over a mold. Many heavy-gauge forming applications use vacuum only in the form process, although some use two halves of mating form tooling and include air pressure to help form. Aircraft windscreens and machine gun turret windows spurred the advance of heavy-gauge forming technology during World War II. Heavy-gauge parts are used as cosmetic surfaces on permanent structures such as kiosks, automobiles, trucks, medical equipment, material handling equipment, refrigerators, spas, and shower enclosures, and electrical and electronic equipment. Unlike most thin-gauge thermoformed parts, heavy-gauge parts are often hand-worked after forming for trimming to final shape or for additional drilling, cutting, or finishing, depending on the product. Heavy-gauge products typically are of a "permanent" end use nature, while thin-gauge parts are more often designed to be disposable or recyclable and are primarily used to package or contain a food item or product. Heavy-gauge thermoforming is typically used for production quantities of 250 to 3000 annually, with lower tooling costs and faster product development than competing plastic technologies like injection molding.

Engineering

Thermoforming has benefited from applications of engineering technology, although the basic forming process is very similar to what was invented many years ago. Microprocessor and computer controls on more modern machinery allow for greatly increased process control and repeatability of same-job setups from one production run with the ability to save oven heater and process timing settings between jobs. The ability to place formed sheet into an inline trim station for more precise trim registration has been hugely improved due to the common use of electric servo motors for chain indexing versus air cylinders, gear racks, and clutches on older machines. Electric servo motors are also used on some modern and more sophisticated forming machines for actuation of the machine platens where form and trim tooling are mounted, rather than air cylinders which have traditionally been the industry standard, giving more precise control over closing and opening speeds and timing of the tooling. Quartz and radiant-panel oven heaters generally provide more precise and thorough sheet heating over older cal-rod type heaters, and better allow for zoning of ovens into areas of adjustable heat.

A new technology, ToolVu, has been developed to provide real-time feedback on thermoformer machines. This stand-alone system connects directly to the thermoformer and utilizes multiple sensors to record production-run data in real time including air pressure, temperature, tool strain gauge and other specifications. The system sends out multiple warnings and alerts whenever pre-set production parameters are compromised during a run. This reduces machine down time, lowers startup time and decreases startup scrap.

An integral part of the thermoforming process is the tooling, which is specific to each part that is to be produced. Thin-gauge thermoforming as described above is almost always performed on in-line machines and typically requires molds, plug assists, pressure boxes and all mounting plates as well as the trim tooling and stacker parts that pertain to the job. Thick or heavy-gauge thermoforming also requires tooling specific to each part, but because the part size can be very large, the molds can be cast aluminum or some other composite material as well as machined aluminum as in thin gauge. Typically, thick-gauge parts must be trimmed on CNC routers or hand trimmed using saws or hand routers. Even the most sophisticated thermoforming machine is limited to the quality of the tooling. Some large thermoforming manufacturers choose to have design and tool making facilities in house while others will rely on outside tool-making shops to build the tooling.

Types of Molds

- Plaster of paris mold,
- Wooden mold,
- Plastic mold,
- Aluminium mold.

LAMINATING

Laminating is the process, or technique of taking a material, or materials, and manufacturing it in order to add multiple layers so that the composite (originally used) materials can achieve improved strength and stability, as well as insulation and appearance, or other properties from the use of different materials. A perfect and permanent lamination assemblage can be accomplished through the use of heat or thermal discharge, significant pressure, welding, and strong adhesives.

Regarding the lamination process, there are different methods that were previously touched on that can be used, depending on the materials that need to be laminated. It is important to consult with a reputable company, such as Jessup Manufacturing, if you want to laminate a particular material or object, but you are unsure of the correct process. The most common laminated objects are paper products – especially for photos, but a myriad number of objects can be laminated, including, glass (windshields) and fiberglass, wood (wooden beams from timber) and plywood, as well as several other materials.

The most popular reason to laminate an object is to add properties to the surface of a material. For example, when laminating printed paper, people are now able to add a magnet to the surface, so they can hang the object on a magnetized object, like a refrigerator. This is definitely one of the more popular reasons to laminate paper materials, as one would probably be hard pressed to find a household anywhere in the United States that does not have something hanging from the refrigerator.

METAL INJECTION MOLDING

Metal injection molding (MIM) is a metalworking process in which finely-powdered metal is mixed with binder material to create a "feedstock" that is then shaped and solidified using injection molding. The molding process allows high volume, complex parts to be shaped in a single step. After molding, the part undergoes conditioning operations to remove the binder (debinding) and densify the powders. Finished products are small components used in many industries and applications.

The behavior of MIM feedstock is governed by rheology, the study of sludges, suspensions, and other non-Newtonian fluids.

Due to current equipment limitations, products must be molded using quantities of 100 grams or less per "shot" into the mold. This shot can be distributed into multiple cavities, making MIM cost-effective for small, intricate, high-volume products, which would otherwise be expensive to produce. MIM feedstock can be composed of a plethora of metals, but most common are stainless steels, widely used in powder metallurgy. After the initial molding, the feedstock binder is removed, and the metal particles are diffusion bonded and densified to achieve the desired strength properties. The latter operation typically shrinks the product by 15% in each dimension.

Eyeglass components produced by powder injection molding.

The metal injection molding market has grown from $9 million USD in 1986, to $382 million USD in 2004 to more than $1.5 billion USD in 2015. A related technology is ceramic powder injection molding, leading to about $2 billion USD total sales. Most of the growth in recent years has been in Asia.

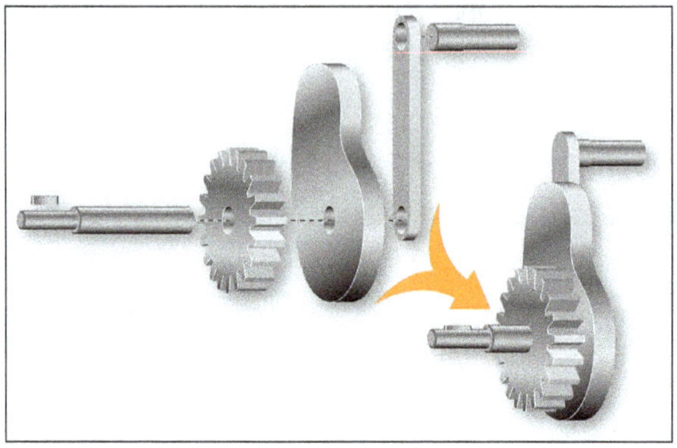

Complex geometry in one component instead of an assembly of many items.

Process

In the monograph P.O. Gribovsky, published in 1956, describes in detail the technology of hot casting (hot molding) ceramic products under pressure (now, Low Pressure Powder Injection Molding) and, in particular, notes that "hot casting technology provides the ability to manufacture products from any solid materials, ranging from natural minerals, pure oxides, carbides, metals, etc., and ending with multicomponent composite synthetic materials and their combinations". This indication of the possibility of MIM-casting, which was implemented by Dr. Raymond E. Wiech Jr. in the 1970s, who refined MIM technology as co-founder of a California company named Parmatech, the name being condensed from the phrase "particle materials technology". Wiech later patented his process, and it was widely adopted for manufacturing use in the 1980s.

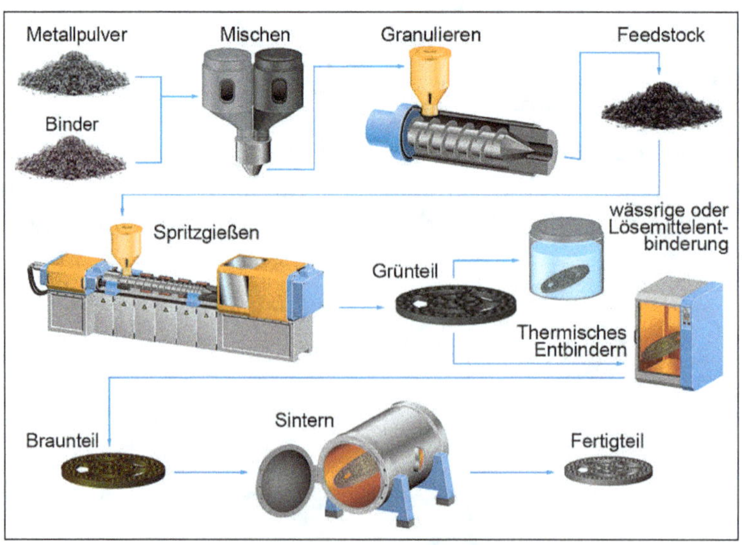

The powder injection molding process.

MIM gained recognition throughout the 1990s as improvements to subsequent conditioning processes resulted in an end product that performs similarly to or better than those made through competing processes. MIM technology improved cost efficiency through high volume production to "net-shape", negating costly, additional operations such as machining although MIM is weak in terms of tight dimensional specifications.

The process steps involve combining metal powders with polymers such as wax and polypropylene binders to produce the "feedstock" mix that is injected as a liquid into a mold using plastic injection molding machines. The molded or "green part" is cooled and ejected from the mold. Next, a portion of the binder material is removed using solvent, thermal furnaces, catalytic process, or a combination of methods. The resulting, fragile and porous (40 volume percent "air") part, is in a condition called the "brown" stage. To improve handling often the debinding and sintering are combined into a single process. Sintering heats the powder to temperatures near the melting point in a protective atmosphere furnace to densify the particles using capillary forces in a process called sintering. MIM parts are often sintered at temperatures nearly high enough to induce partial melting in a process termed liquid phase sintering. For example, a stainless steel might be heated to 1350 to 1400 degrees Celsius). Diffusion rates are high leading to high shrinkage and densification. If performed in vacuum, it is common to reach 96–99% solid density. The end-product metal has comparable mechanical and physical properties with annealed parts made using classic metalworking methods. Post sintering heat treatments for MIM are the same as with other fabrication routes, and with high density the MIM component is compatible with the metal conditioning treatments such as plating, passivating, annealing, carburizing, nitriding, and precipitation hardening.

Applications

The window of economic advantage in metal injection molded parts lies in complexity and volume for small-size parts. MIM materials are comparable to metal formed by competing methods, and final products are used in a broad range of industrial, commercial, medical, dental, firearms, aerospace, and automotive applications. Dimensional tolerances of ±0.3% are common and machining is required for closer tolerances. MIM can produce parts where it is difficult, or even impossible, to efficiently manufacture an item through other means of fabrication. Ideally, at least 75 dimensional specifications in a component of just 25 mm maximum size and 10 g mass is best - as for example required for watch cases, cellular telephone plugs, and laptop computer hinges. Increased costs for traditional manufacturing methods inherent to part complexity, such as internal/external threads, miniaturization, or identity marking, typically do not increase the cost in a MIM operation due to the flexibility of injection molding.

Other design capabilities that can be implemented into the MIM operation include product codes, part numbers, or date stamps; parts manufactured to their net weight reducing material waste and cost; Density controlled to within 95–98%; Amalgamation of parts and Complex 3D Geometries.

The ability to combine several operations into one process ensures MIM is successful in saving lead times as well as costs, providing significant benefits to manufacturers. The metal injection molding process might be a green technology due to the significant reduction in wastage compared to "traditional" manufacturing methods such as 5 axis CNC machining. However, some of the older operations generate toxic emissions such as formaldehyde, dispose of chlorinated solvents, and must burn off wax or other polymers, leading to greenhouse gas emissions.

There is a broad range of materials available when utilizing the MIM process. Traditional metalworking processes often involve a significant amount of material waste, which makes MIM a highly efficient option for the fabrication of complex components consisting of expensive/special alloys (cobalt-chrome, 17-4 PH stainless steel, titanium alloys and tungsten carbides). MIM is a viable option when extremely thin walls specifications (i.e., 100 micrometers) are required. Additionally, EMI shielding (Electromagnetic Interference) requirements has presented unique challenges, which are being successfully attained through the utilization of specialty alloys (ASTM A753 Type 4).

TRANSFER MOLDING

Transfer molding (BrE moulding) is a manufacturing process where casting material is forced into a mold. Transfer molding is different from compression molding in that the mold is enclosed rather than open to the fill plunger resulting in higher dimensional tolerances and less environmental impact. Compared to injection molding, transfer molding uses higher pressures to uniformly fill the mold cavity. This allows thicker reinforcing fiber matrices to be more completely saturated by resin. Furthermore, unlike injection molding the transfer mold casting material may start the process as a solid. This can reduce equipment costs and time dependency. The transfer process may have a slower fill rate than an equivalent injection molding processes.

Process

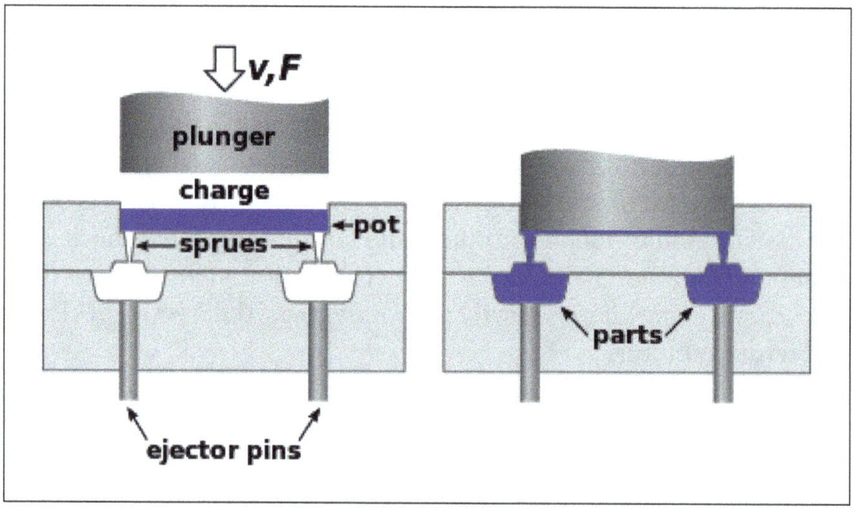

Transfer Molding basic process.

The mold interior surfaces may be gel-coated. If desired the mold is first pre-loaded with a reinforcing fiber matrix or preform. Fiber content of a transfer molded composite can be as high as 60% by volume. The fill material may be a preheated solid or a liquid. It is loaded into a chamber known as the pot. A ram or plunger forces material from the pot into the heated mold cavity. If feed-stock is initially solid, the forcing pressure and mold temperature melt it. Standard mold features such as sprue channels, a flow gate and ejector pins may be used. The heated mold ensures that the flow remains liquid for complete filling. Once filled the mold can be cooled at a controlled rate for optimal thermoset curing.

Variations

The industry identifies a variety of processes within the transfer molding category. There are areas of overlap and the distinctions between each method may not be clearly defined.

Resin Transfer Molding

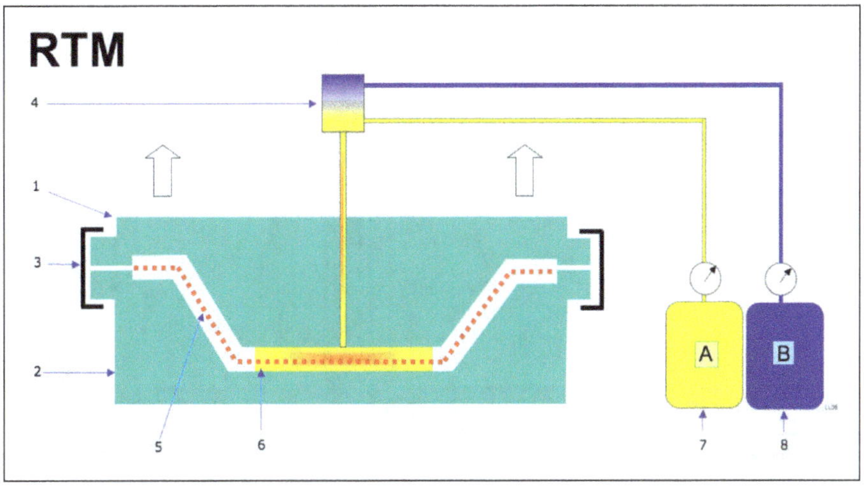

Resin Transfer Molding 1: Cope 2: Drag 3: Clamp 4: Mixing chamber 5: Fiber preform 6: Heated mold 7: Resin 8: Curative.

Resin transfer molding (RTM) uses a liquid thermoset resin to saturate a fiber preform placed in a closed mold. The process is versatile and can fabricate products with embedded objects such as foam cores or other components in addition to the fiber preform.

Vacuum Assisted Resin Transfer Molding

Vacuum assisted transfer molding (VARTM) uses a partial vacuum on one side of a fiber mat to pull the resin in for complete saturation. VARTM uses lower plunger forces which allows molding to be carried out with cheaper equipment. The use of a vacuum may allow the resin to adequately flow and or cure without heating. This temperature independence allows thicker fiber preforms and larger product geometries to be economical. VARTM can produce parts with less porosity than regular transfer molding with a proportional increase in casting strength.

Micro Transfer Molding

Also called transfer micromolding, this is a process that uses a mold to form then transfer structures as small as 30 nm onto thin films and microcircuitry. Unlike normal scale transfer molding, the micro form can and is used with metals as well as non metals.

Defects

Limiting defects is key when commercial producing any sort of material. Transfer molding is no exception. For example, voids in a transfer molded parts significantly reduce strength and modulus. There can also be defects when fibers are used around sharp corners. The resin flow can create resin rich zones on the outside of these corners.

Pressure Distribution

There are several contributing factors to voids in the final product of transfer molding. One is a non uniform pressure distribution among the material being pressed into the mold. In this case the material folds in on itself and generates voids. Another is voids in the resin being forced into the mold beforehand. This may be obvious, but it is a main contributor. Things to be done to limit these molds include pressing the resin in at a high pressure, keeping the fiber distribution uniform, and using a high quality properly degassed base resin.

Sharp Corners

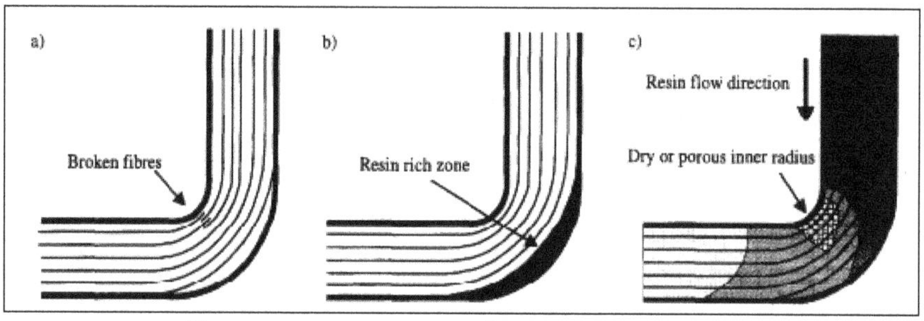

Sharp corner generates voids in transfer molding.

Sharp corners are the problems with all mold based manufacturing, including casting. Specifically in transfer molding corners can break fibers that have been placed in the mold and can create voids on the inside of corners. This effect is demonstrated in Figure on the right. The limiting factor in these designs is the inner corner radius. This inner radius limit varies depending on resin and fiber selection, but a rule of thumbs is the radius though be 3 to 5 times the laminate thickness.

Materials

The material most commonly used for transfer molding is a thermoset polymer. This type of polymer is easy to mold and manipulate, but upon curing, hardens into a permanent form. For simple homogeneous transfer molded parts, the part is simply made of this plastic substrate. On the other hand, resin transfer molding allows for a composite material to be made by placing a fiber within the mold and subsequently injecting the thermosetting polymer.

Defects known as voids and dry resin (in the case of resin transfer molding) are possible in transfer molding and often are exacerbated by high viscosity materials. This is because a high viscosity plastic flowing through a thin mold may miss entire vacated areas, leaving air pockets. When air pockets are left in the presence of fiber, this creates a "dry" area, which prevents load from being transferred through the fibers in the dry area.

Materials used for the plastic are often polyurethanes or epoxy resins. Both of these are soft and malleable before curing, becoming much harder after setting. Materials used for fibers vary extensively, although common choices are carbon or Kevlar fibers, as well as organic fibers, such as hemp.

INJECTION MOLDING

Injection moulding is a manufacturing process for producing parts by injecting molten material into a mould. Injection moulding can be performed with a host of materials mainly including metals (for which the process is called die-casting), glasses, elastomers, confections, and most commonly thermoplastic and thermosetting polymers. Material for the part is fed into a heated barrel, mixed (Using a helical shaped screw), and injected (Forced) into a mould cavity, where it cools and hardens to the configuration of the cavity. After a product is designed, usually by an industrial designer or an engineer, moulds are made by a mould-maker (or toolmaker) from metal, usually either steel or aluminium, and precision-machined to form the features of the desired part. Injection moulding is widely used for manufacturing a variety of parts, from the smallest components to entire body panels of cars. Advances in 3D printing technology, using photopolymers which do not melt during the injection moulding of some lower temperature thermoplastics, can be used for some simple injection moulds.

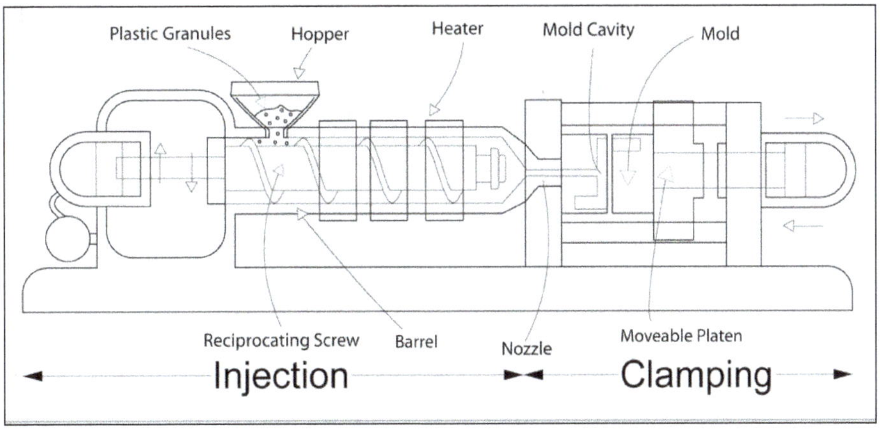

An injection moulding machine.

Parts to be injection moulded must be very carefully designed to facilitate the moulding process; the material used for the part, the desired shape and features of the part, the material of the mould, and the properties of the moulding machine must all be taken into account. The versatility of injection moulding is facilitated by this breadth of design considerations and possibilities.

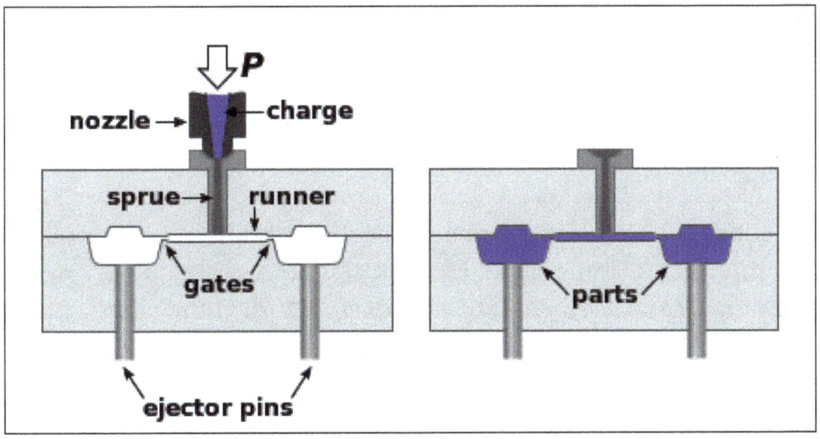

Simplified diagram of the process.

Applications

Injection moulding is used to create many things such as wire spools, packaging, bottle caps, automotive parts and components, toys, pocket combs, some musical instruments (and parts of them), one-piece chairs and small tables, storage containers, mechanical parts (including gears), and most other plastic products available today. Injection moulding is the most common modern method of manufacturing plastic parts; it is ideal for producing high volumes of the same object.

Process Characteristics

Thermoplastic resin pellets for injection moulding.

Injection moulding uses a ram or screw-type plunger to force molten plastic material into a mould cavity; this solidifies into a shape that has conformed to the contour of the mould. It is most commonly used to process both thermoplastic and thermosetting polymers, with the volume used of the former being considerably higher. Thermoplastics are prevalent due to characteristics which make them highly suitable for injection moulding, such as the ease with which they may be recycled, their versatility allowing them to be used in a wide variety of applications, and their ability to soften and flow upon heating. Thermoplastics also have an element of safety over thermosets; if a thermosetting polymer is not ejected from the injection barrel in a timely manner, chemical crosslinking may occur causing the screw and check valves to seize and potentially damaging the injection moulding machine.

Injection moulding consists of the high pressure injection of the raw material into a mould which shapes the polymer into the desired shape. Moulds can be of a single cavity or multiple cavities. In multiple cavity moulds, each cavity can be identical and form the same parts or can be unique and form multiple different geometries during a single cycle. Moulds are generally made from tool steels, but stainless steels and aluminium moulds are suitable for certain applications. Aluminium moulds are typically ill-suited for high volume production or parts with narrow dimensional tolerances, as they have inferior mechanical properties and are more prone to wear, damage, and deformation during the injection and clamping cycles; however, aluminium moulds are cost-effective in low-volume applications, as mould fabrication costs and time are considerably reduced. Many steel moulds are designed to process well over a million parts during their lifetime and can cost hundreds of thousands of dollars to fabricate.

When thermoplastics are molded, typically pelletised raw material is fed through a hopper into a heated barrel with a reciprocating screw. Upon entrance to the barrel, the temperature increases and the Van der Waals forces that resist relative flow of individual chains are weakened as a result of increased space between molecules at higher thermal energy states. This process reduces its viscosity, which enables the polymer to flow with the driving force of the injection unit. The screw delivers the raw material forward, mixes and homogenises the thermal and viscous distributions of the polymer, and reduces the required heating time by mechanically shearing the material and adding a significant amount of frictional heating to the polymer. The material feeds forward through a check valve and collects at the front of the screw into a volume known as a shot. A shot is the volume of material that is used to fill the mould cavity, compensate for shrinkage, and provide a cushion (approximately 10% of the total shot volume, which remains in the barrel and prevents the screw from bottoming out) to transfer pressure from the screw to the mould cavity. When enough material has gathered, the material is forced at high pressure and velocity into the part forming cavity. The exact amount of shrinkage is a function of the resin being used, and can be relatively predictable. To prevent spikes in pressure, the process normally uses a transfer position corresponding to a 95–98% full cavity where the screw shifts from a constant velocity to a constant pressure control. Often injection times are well under 1 second. Once the screw reaches the transfer position the packing pressure is applied, which completes mould filling and compensates for thermal shrinkage, which is quite high for thermoplastics relative to many other materials. The packing pressure is applied until the gate (cavity entrance) solidifies. Due to its small size, the gate is normally the first place to solidify through its entire thickness. Once the gate solidifies, no more material can enter the cavity; accordingly, the screw reciprocates and acquires material for the next cycle while the material within the mould cools so that it can be ejected and be dimensionally stable. This cooling duration is dramatically reduced by the use of cooling lines circulating water or oil from an external temperature controller. Once the required temperature has been achieved, the mould opens and an array of pins, sleeves, strippers, etc. are driven forward to demould the article. Then, the mould closes and the process is repeated.

For a two shot mould, two separate materials are incorporated into one part. This type of injection moulding is used to add a soft touch to knobs, to give a product multiple colours, or to produce a part with multiple performance characteristics.

For thermosets, typically two different chemical components are injected into the barrel. These components immediately begin irreversible chemical reactions which eventually crosslinks the material into a single connected network of molecules. As the chemical reaction occurs, the two fluid components permanently transform into a viscoelastic solid. Solidification in the injection barrel and screw can be problematic and have financial repercussions; therefore, minimising the thermoset curing within the barrel is vital. This typically means that the residence time and temperature of the chemical precursors are minimised in the injection unit. The residence time can be reduced by minimising the barrel's volume capacity and by maximising the cycle times. These factors have led to the use of a thermally isolated, cold injection unit that injects the reacting chemicals into a thermally isolated hot mould, which increases the rate of chemical reactions and results in shorter time required to achieve a solidified thermoset component. After the part has solidified, valves close to isolate the injection system and chemical precursors, and the mould opens to eject the moulded parts. Then, the mould closes and the process repeats.

Pre-moulded or machined components can be inserted into the cavity while the mould is open, allowing the material injected in the next cycle to form and solidify around them. This process is known as Insert moulding and allows single parts to contain multiple materials. This process is often used to create plastic parts with protruding metal screws, allowing them to be fastened and unfastened repeatedly. This technique can also be used for In-mould labelling and film lids may also be attached to moulded plastic containers.

A parting line, sprue, gate marks, and ejector pin marks are usually present on the final part. None of these features are typically desired, but are unavoidable due to the nature of the process. Gate marks occur at the gate which joins the melt-delivery channels (sprue and runner) to the part forming cavity. Parting line and ejector pin marks result from minute misalignments, wear, gaseous vents, clearances for adjacent parts in relative motion, and/or dimensional differences of the mating surfaces contacting the injected polymer. Dimensional differences can be attributed to non-uniform, pressure-induced deformation during injection, machining tolerances, and non-uniform thermal expansion and contraction of mould components, which experience rapid cycling during the injection, packing, cooling, and ejection phases of the process. Mould components are often designed with materials of various coefficients of thermal expansion. These factors cannot be simultaneously accounted for without astronomical increases in the cost of design, fabrication, processing, and quality monitoring. The skillful mould and part designer will position these aesthetic detriments in hidden areas if feasible.

Polymers Best Suited for the Process

Most polymers, sometimes referred to as resins, may be used, including all thermoplastics, some thermosets, and some elastomers. Since 1995, the total number of available materials for injection moulding has increased at a rate of 750 per year; there were approximately 18,000 materials available when that trend began. Available materials include alloys or blends of previously developed materials, so product designers can choose the material with the best set of properties from a vast selection. Major criteria for selection of a material are the strength and function required for the final part, as well as the cost, but also each material has different parameters for moulding that must be taken into account. Other considerations when choosing an injection molding material include flexural modulus of elasticity, or the degree to which a material can be bent without damage, as well as heat deflection and water absorption. Common polymers like epoxy and phenolic are examples of thermosetting plastics while nylon, polyethylene, and polystyrene are thermoplastic. Until comparatively recently, plastic springs were not possible, but advances in polymer properties make them now quite practical. Applications include buckles for anchoring and disconnecting the outdoor-equipment webbing.

Equipment

Injection moulding machines consist of a material hopper, an injection ram or screw-type plunger, and a heating unit. Also known as platens, they hold the moulds in which the components are shaped. Presses are rated by tonnage, which expresses the amount of clamping force that the machine can exert. This force keeps the mould closed during the injection process. Tonnage can vary from less than 5 tons to over 9,000 tons, with the higher figures used in comparatively few manufacturing operations. The total clamp force needed is determined by the projected area of the part being moulded. This projected area is multiplied by a clamp force of from 1.8 to 7.2 tons for each square centimetre of the projected areas. As a rule of thumb, 4 or 5 tons/in^2 can be used for most products. If the plastic material is very stiff, it will require more injection pressure to fill the mould, and thus more clamp tonnage to hold the mould closed. The required force can also be determined by the material used and the size of the part. Larger parts require higher clamping force.

Paper clip mould opened in moulding machine; the nozzle is visible at right.

Mould

Mould or die are the common terms used to describe the tool used to produce plastic parts in moulding.

Since moulds have been expensive to manufacture, they were usually only used in mass production where thousands of parts were being produced. Typical moulds are constructed from hardened steel, pre-hardened steel, aluminium, and/or beryllium-copper alloy. The choice of material to build a mould from is primarily one of economics; in general, steel moulds cost more to construct, but their longer lifespan will offset the higher initial cost over a higher number of parts made before wearing out. Pre-hardened steel moulds are less wear-resistant and are used for lower volume requirements or larger components; their typical steel hardness is 38–45 on the Rockwell-C scale. Hardened steel moulds are heat treated after machining; these are by far superior in terms of wear resistance and lifespan. Typical hardness ranges between 50 and 60 Rockwell-C (HRC). Aluminium moulds can cost substantially less, and when designed and machined with modern computerised equipment can be economical for moulding tens or even hundreds of thousands of parts. Beryllium copper is used in areas of the mould that require fast heat removal or areas that see the most shear heat generated. The moulds can be manufactured either by CNC machining or by using electrical discharge machining processes.

Injection Moulding Die with Side Pulls

Mould Design

The mould consists of two primary components, the injection mould (A plate) and the ejector mould (B plate). These components are also referred to as moulder and mouldmaker. Plastic resin enters the mould through a sprue or gate in the injection mould; the sprue bushing is to seal tightly against the nozzle of the injection barrel of the moulding machine and to allow molten plastic to flow from the barrel into the mould, also known as the cavity. The sprue bushing directs the molten plastic to the cavity images through channels that are machined into the faces of the A and B plates. These channels allow plastic to run along them, so they are referred to as runners. The molten plastic flows through the runner and enters one or more specialised gates and into the cavity geometry to form the desired part.

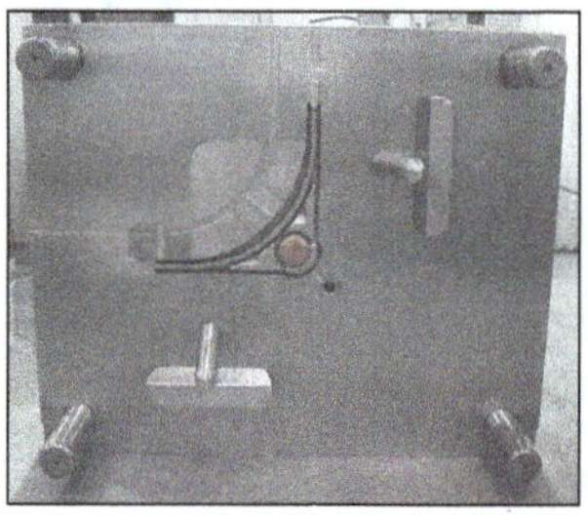

"A" side of die for 25% glass-filled acetal with 2 side pulls.

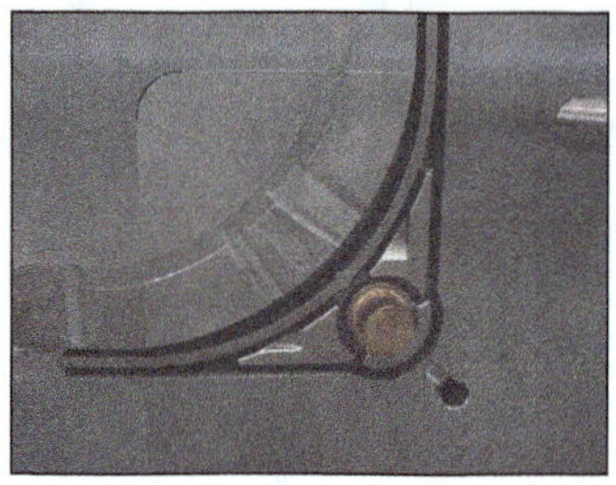

Close up of removable insert in "A" side.

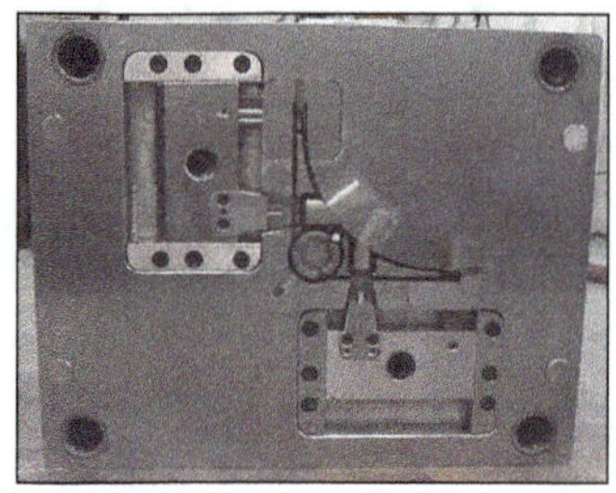

"B" side of die with side pull actuators.

Insert removed from die.

Standard two plates tooling – core and cavity are inserts in a mould base – "family mould" of five different parts.

The amount of resin required to fill the sprue, runner and cavities of a mould comprises a "shot". Trapped air in the mould can escape through air vents that are ground into the parting line of the mould, or around ejector pins and slides that are slightly smaller than the holes retaining them. If the trapped air is not allowed to escape, it is compressed by the pressure of the incoming material and squeezed into the corners of the cavity, where it prevents filling and can also cause other defects. The air can even become so compressed that it ignites and burns the surrounding plastic material.

To allow for removal of the moulded part from the mould, the mould features must not overhang one another in the direction that the mould opens, unless parts of the mould are designed to move from between such overhangs when the mould opens (using components called Lifters).

Sides of the part that appear parallel with the direction of draw (the axis of the cored position (hole) or insert is parallel to the up and down movement of the mould as it opens and closes) are typically angled slightly, called draft, to ease release of the part from the mould. Insufficient draft can cause deformation or damage. The draft required for mould release is primarily dependent on the depth of the cavity; the deeper the cavity, the more draft necessary. Shrinkage must also be taken into account when determining the draft required. If the skin is too thin, then the moulded part will tend to shrink onto the cores that form while cooling and cling to those cores, or the part may warp, twist, blister or crack when the cavity is pulled away.

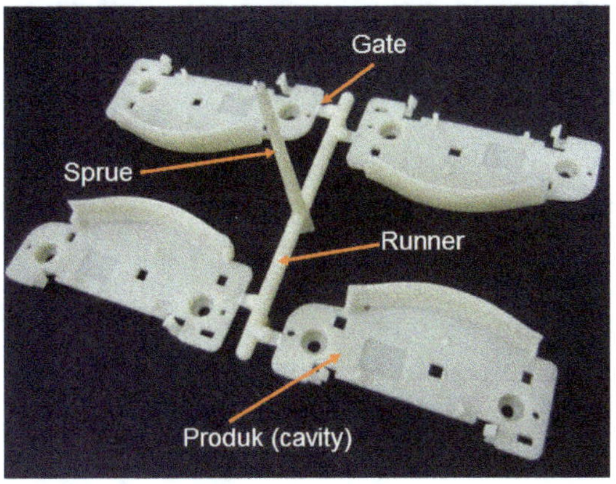

Sprue, runner and gates in actual injection moulding product.

A mould is usually designed so that the moulded part reliably remains on the ejector (B) side of the mould when it opens, and draws the runner and the sprue out of the (A) side along with the parts. The part then falls freely when ejected from the (B) side. Tunnel gates, also known as submarine or mould gates, are located below the parting line or mould surface. An opening is machined into the surface of the mould on the parting line. The moulded part is cut (by the mould) from the runner system on ejection from the mould. Ejector pins, also known as knockout pins, are circular pins placed in either half of the mould (usually the ejector half), which push the finished moulded product, or runner system out of a mould. The ejection of the article using pins, sleeves, strippers, etc., may cause undesirable impressions or distortion, so care must be taken when designing the mould.

The standard method of cooling is passing a coolant (usually water) through a series of holes drilled through the mould plates and connected by hoses to form a continuous pathway. The coolant absorbs heat from the mould (which has absorbed heat from the hot plastic) and keeps the mould at a proper temperature to solidify the plastic at the most efficient rate.

To ease maintenance and venting, cavities and cores are divided into pieces, called inserts, and sub-assemblies, also called inserts, blocks, or chase blocks. By substituting interchangeable inserts, one mould may make several variations of the same part.

More complex parts are formed using more complex moulds. These may have sections called slides, that move into a cavity perpendicular to the draw direction, to form overhanging part features. When the mould is opened, the slides are pulled away from the plastic part by using stationary "angle pins" on the stationary mould half. These pins enter a slot in the slides and cause the slides to move backward when the moving half of the mould opens. The part is then ejected and the mould closes. The closing action of the mould causes the slides to move forward along the angle pins.

Some moulds allow previously moulded parts to be reinserted to allow a new plastic layer to form around the first part. This is often referred to as overmoulding. This system can allow for production of one-piece tires and wheels.

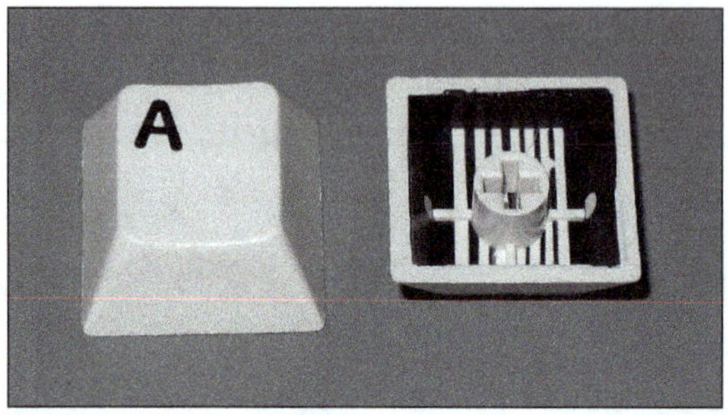

Two-shot injection moulded keycaps from a computer keyboard.

Two-shot or multi-shot moulds are designed to "overmould" within a single moulding cycle and must be processed on specialised injection moulding machines with two or more injection units. This process is actually an injection moulding process performed twice and therefore has a much smaller margin of error. In the first step, the base colour material is moulded into a basic shape, which contains spaces for the second shot. Then the second material, a different colour, is injection-moulded into those spaces. Pushbuttons and keys, for instance, made by this process have markings that cannot wear off, and remain legible with heavy use.

A mould can produce several copies of the same parts in a single "shot". The number of "impressions" in the mould of that part is often incorrectly referred to as cavitation. A tool with one impression will often be called a single impression (cavity) mould. A mould with 2 or more cavities of the same parts will likely be referred to as multiple impression (cavity) mould. Some extremely high production volume moulds (like those for bottle caps) can have over 128 cavities.

In some cases, multiple cavity tooling will mould a series of different parts in the same tool. Some toolmakers call these moulds family moulds as all the parts are related. Some examples include plastic model kits.

Mould Storage

Manufacturers go to great lengths to protect custom moulds due to their high average costs. The perfect temperature and humidity level is maintained to ensure the longest possible lifespan for each custom mould. Custom moulds, such as those used for rubber injection moulding, are stored in temperature and humidity controlled environments to prevent warping.

Tool Materials

Beryllium-copper insert (yellow) on injection moulding mould for ABS resin.

Tool steel is often used. Mild steel, aluminium, nickel or epoxy are suitable only for prototype or very short production runs. Modern hard aluminium (7075 and 2024 alloys) with proper mould design, can easily make moulds capable of 100,000 or more part life with proper mould maintenance.

Machining

Moulds are built through two main methods: standard machining and EDM. Standard machining, in its conventional form, has historically been the method of building injection moulds. With technological developments, CNC machining became the predominant means of making more complex moulds with more accurate mould details in less time than traditional methods.

The electrical discharge machining (EDM) or spark erosion process has become widely used in mould making. As well as allowing the formation of shapes that are difficult to machine, the process allows pre-hardened moulds to be shaped so that no heat treatment is required. Changes to a hardened mould by conventional drilling and milling normally require annealing to soften the mould, followed by heat treatment to harden it again. EDM is a simple process in which a shaped electrode, usually made of copper or graphite, is very slowly lowered onto the mould surface (over a period of many hours), which is immersed in paraffin oil (kerosene). A voltage applied between tool and mould causes spark erosion of the mould surface in the inverse shape of the electrode.

Cost

The number of cavities incorporated into a mould will directly correlate in moulding costs. Fewer cavities require far less tooling work, so limiting the number of cavities in-turn will result in lower initial manufacturing costs to build an injection mould.

As the number of cavities play a vital role in moulding costs, so does the complexity of the part's design. Complexity can be incorporated into many factors such as surface finishing, tolerance requirements, internal or external threads, fine detailing or the number of undercuts that may be incorporated.

Further details, such as undercuts or any feature causing additional tooling, will increase the mould cost. Surface finish of the core and cavity of moulds will further influence the cost.

Rubber injection moulding process produces a high yield of durable products, making it the most efficient and cost-effective method of moulding. Consistent vulcanisation processes involving precise temperature control significantly reduces all waste material.

Injection Process

With injection moulding, granular plastic is fed by a forced ram from a hopper into a heated barrel. As the granules are slowly moved forward by a screw-type plunger, the plastic is forced into a heated chamber, where it is melted. As the plunger advances, the melted plastic is forced through a nozzle that rests against the mould, allowing it to enter the mould cavity through a gate and runner system. The mould remains cold so the plastic solidifies almost as soon as the mould is filled.

Small injection moulder showing hopper, nozzle and die area.

Injection Moulding Cycle

The sequence of events during the injection mould of a plastic part is called the injection moulding cycle. The cycle begins when the mould closes, followed by the injection of the polymer into the mould cavity. Once the cavity is filled, a holding pressure is maintained to compensate for material shrinkage. In the next step, the screw turns, feeding the next shot to the front screw. This causes the screw to retract as the next shot is prepared. Once the part is sufficiently cool, the mould opens and the part is ejected.

Scienti ic versus Traditional Moulding

Traditionally, the injection portion of the moulding process was done at one constant pressure to fill and pack the cavity. This method, however, allowed for a large variation in dimensions from cycle-to-cycle. More commonly used now is scientific or decoupled moulding, a method pioneered by RJG Inc. In this the injection of the plastic is "decoupled" into stages to allow better control of part dimensions and more cycle-to-cycle (commonly called shot-to-shot in the industry) consistency. First the cavity is filled to approximately 98% full using velocity (speed) control. Although the pressure should be sufficient to allow for the desired speed, pressure limitations during this stage are undesirable. Once the cavity is 98% full, the machine switches from velocity control to pressure control, where the cavity is "packed out" at a constant pressure, where sufficient velocity to reach desired pressures is required. This allows part dimensions to be controlled to within thousandths of an inch or better.

Different Types of Injection Moulding Processes

Although most injection moulding processes are covered by the conventional process description above, there are several important moulding variations including, but not limited to:

- Die casting.
- Metal injection moulding.
- Thin-wall injection moulding.
- Injection moulding of liquid silicone rubber.
- Reaction injection moulding.

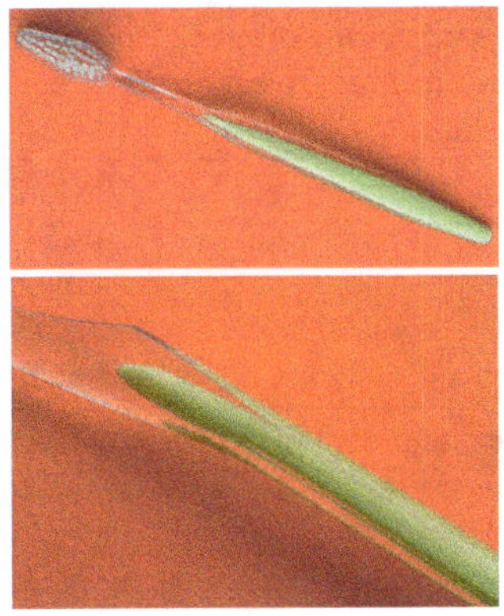

Sandwich-moulded toothbrush handle.

Process Troubleshooting

Like all industrial processes, injection moulding can produce flawed parts. In the field of injection moulding, troubleshooting is often performed by examining defective parts for specific defects and addressing these defects with the design of the mould or the characteristics of the process itself. Trials are often performed before full production runs in an effort to predict defects and determine the appropriate specifications to use in the injection process.

When filling a new or unfamiliar mould for the first time, where shot size for that mould is unknown, a technician/tool setter may perform a trial run before a full production run. They start with a small shot weight and fills gradually until the mould is 95 to 99% full. Once this is achieved, a small amount of holding pressure will be applied and holding time increased until gate freeze off (solidification time) has occurred. Gate freeze off time can be determined by increasing the hold time, and then weighing the part. When the weight of the part does not change, it is then known that the gate has frozen and no more material is injected into the part. Gate solidification time is important, as this determines cycle time and the quality and consistency of the product, which itself is an important issue in the economics of the production process. Holding pressure is increased until the parts are free of sinks and part weight has been achieved.

Moulding Defects

Injection moulding is a complex technology with possible production problems. They can be caused either by defects in the moulds, or more often by the moulding process itself.

Moulding defects	Alternative name	Descriptions	Causes
Blister	Blistering	Raised or layered zone on surface of the part.	Tool or material is too hot, often caused by a lack of cooling around the tool or a faulty heater.
Burn marks	Air burn/gas burn/dieseling/ gas marks/Blow marks.	Black or brown burnt areas on the part located at furthest points from gate or where air is trapped.	Tool lacks venting, injection speed is too high.
Color streaks (US)	Colour streaks (UK)	Localised change of colour.	Masterbatch isn't mixing properly, or the material has run out and it's starting to come through as natural only. Previous coloured material "dragging" in nozzle or check valve.

Contamination	Unwanted or foreign material.	Different colour matter seen in product, weakening the product.	Poor material introduced by bad recycling or regrind policy; may include floor sweepings, dust and debris.
Delamination		Thin mica like layers formed in part wall.	Contamination of the material e.g. PP mixed with ABS, very dangerous if the part is being used for a safety critical application as the material has very little strength when delaminated as the materials cannot bond.
Flash		Excess material in thin layer exceeding normal part geometry.	Mould is over packed or parting line on the tool is damaged, too much injection speed/material injected, clamping force too low. Can also be caused by dirt and contaminants around tooling surfaces.
Embedded contaminates	Embedded particulates.	Foreign particle (burnt material or other) embedded in the part.	Particles on the tool surface, contaminated material or foreign debris in the barrel, or too much shear heat burning the material prior to injection.
Flow marks	Flow lines.	Directionally "off tone" wavy lines or patterns.	Injection speeds too slow (the plastic has cooled down too much during injection, injection speeds should be set as fast as is appropriate for the process and material used).
Gate Blush	Halo or Blush Marks.	Circular pattern around gate, normally only an issue on hot runner molds.	Injection speed is too fast, gate/sprue/runner size is too small, or the melt/mold temp is too low.
Jetting		Part deformed by turbulent flow of material.	Poor tool design, gate position or runner. Injection speed set too high. Poor design of gates which cause too little die swell and result jetting.
Knit lines	Weld lines	Small lines on the backside of core pins or windows in parts that look like just lines.	Caused by the melt-front flowing around an object standing proud in a plastic part as well as at the end of fill where the melt-front comes together again. Can be minimised or eliminated with a mould-flow study when the mould is in design phase. Once the mould is made and the gate is placed, one can minimise this flaw only by changing the melt and the mould temperature.
Polymer degradation		Polymer breakdown from hydrolysis, oxidation etc.	Excess water in the granules, excessive temperatures in barrel, excessive screw speeds causing high shear heat, material being allowed to sit in the barrel for too long, too much regrind being used.
Sink marks	[sinks]	Localised depression (In thicker zones).	Holding time/pressure too low, cooling time too short, with sprueless hot runners this can also be caused by the gate temperature being set too high. Excessive material or walls too thick.
Short shot	Short fill or short mould	Partial part	Lack of material, injection speed or pressure too low, mould too cold, lack of gas vents.
Splay marks	Splash mark or silver streaks	Usually appears as silver streaks along the flow pattern, however depending on the type and colour of material it may represent as small bubbles caused by trapped moisture.	Moisture in the material, usually when hygroscopic resins are dried improperly. Trapping of gas in "rib" areas due to excessive injection velocity in these areas. Material too hot, or is being sheared too much.

Stringiness	Stringing or long-gate	String like remnant from previous shot transfer in new shot.	Nozzle temperature too high. Gate hasn't frozen off, no decompression of the screw, no sprue break, poor placement of the heater bands inside the tool.
Voids		Empty space within part (air pocket is commonly used)	Lack of holding pressure (holding pressure is used to pack out the part during the holding time). Filling too fast, not allowing the edges of the part to set up. Also mould may be out of registration (when the two halves don't centre properly and part walls are not the same thickness). The provided information is the common understanding, Correction: The Lack of pack (not holding) pressure (pack pressure is used to pack out even though is the part during the holding time). Filling too fast does not cause this condition, as a void is a sink that did not have a place to happen. In other words, as the part shrinks the resin separated from itself as there was not sufficient resin in the cavity. The void could happen at any area or the part is not limited by the thickness but by the resin flow and thermal conductivity, but it is more likely to happen at thicker areas like ribs or bosses. Additional root causes for voids are un-melt on the melt pool.
Weld line	Knit line/Meld line/Transfer line	Discoloured line where two flow fronts meet.	Mould or material temperatures set too low (the material is cold when they meet, so they don't bond). Time for transition between injection and transfer (to packing and holding) is too early.
Warping	Twisting	Distorted part	Cooling is too short, material is too hot, lack of cooling around the tool, incorrect water temperatures (the parts bow inwards towards the hot side of the tool) Uneven shrinking between areas of the part.
Cracks	Crazing	Improper fusion of two fluid flows, a state before weld line.	Threadline gap in between part due to improper gate location in complex design parts including excess of holes (multipoint gates to be provided), process optimization, proper air venting.

Methods such as industrial CT scanning can help with finding these defects externally as well as internally.

Tolerances

Moulding tolerance is a specified allowance on the deviation in parameters such as dimensions, weights, shapes, or angles, etc. To maximise control in setting tolerances there is usually a minimum and maximum limit on thickness, based on the process used. Injection moulding typically is capable of tolerances equivalent to an IT Grade of about 9–14. The possible tolerance of a thermoplastic or a thermoset is ±0.008 to ±0.002 inches. In specialised applications tolerances as low as ±5 μm on both diameters and linear features are achieved in mass production. Surface finishes of 0.0500 to 0.1000 μm or better can be obtained. Rough or pebbled surfaces are also possible.

Moulding Type	Typical [mm]	Possible [mm]
Thermoplastic	±0.500	±0.200
Thermoset	±0.500	±0.200

Power Requirements

The power required for this process of injection moulding depends on many things and varies between materials used. Manufacturing Processes Reference Guide states that the power requirements depend on "a material's specific gravity, melting point, thermal conductivity, part size, and molding rate."

Material	Specific gravity	Melting point (°F)	Melting point (°C)
Epoxy	1.12 to 1.24	248	120
Phenolic	1.34 to 1.95	248	120
Nylon	1.01 to 1.15	381 to 509	194 to 265
Polyethylene	0.91 to 0.965	230 to 243	110 to 117
Polystyrene	1.04 to 1.07	338	170

Robotic Moulding

Automation means that the smaller size of parts permits a mobile inspection system to examine multiple parts more quickly. In addition to mounting inspection systems on automatic devices, multiple-axis robots can remove parts from the mould and position them for further processes.

Specific instances include removing of parts from the mould immediately after the parts are created, as well as applying machine vision systems. A robot grips the part after the ejector pins have been extended to free the part from the mould. It then moves them into either a holding location or directly onto an inspection system. The choice depends upon the type of product, as well as the general layout of the manufacturing equipment. Vision systems mounted on robots have greatly enhanced quality control for insert moulded parts. A mobile robot can more precisely determine the placement accuracy of the metal component, and inspect faster than a human can.

HOT ISOSTATIC PRESSING

Hot Isostatic Pressing (HIP) is one of material processing methods, which compresses materials by applying high temperature of several hundreds to 2000 °C and isostatic pressure of several tens to 200MPa at the same time. Argon is the most commonly used pressure medium.

Hot pressing is very similar to HIP. Milling, forging, extrusion also apply high temperature and pressure, but not isostatic pressure unlike HIP.

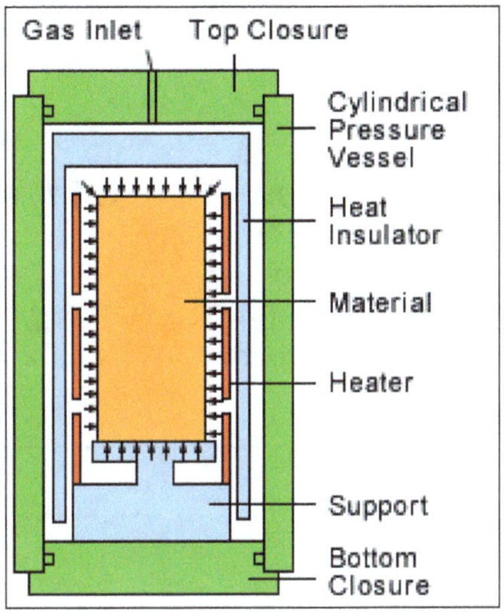

Difference between HIP and Hot Pressing

HIP applies isostatic pressure to materials using gas pressure, while hot pressing applies only uniaxial pressure.

To explain the difference of HIP and hot pressing clearly, suppose that HIP or hot pressing is applied to Material A (metal with pores inside) and Material B (metal with uneven ends). In case of HIP, Material A, as shown in Figure, will contract keeping its initial shape until pores inside disappear, and bond together due to diffusion effects. On the other hand, Material B undergoes no shape change at all because uniform pressure is applied to the uneven edges. In case of hot pressing, the same phenomena as the case of HIP occur to Material A, which is shown in Figure. Material B, however, can't keep its initial uneven shape because pressure is applied only to the convex portions. Both Material A and Material B will have different final shapes after hot pressing depending on shapes of a mold and a punch used. Fabrication of large products and moldings under high temperature might be difficult because of ununiformity due to friction force with a mold and constraints due to temperature and dimensions during the deformation.

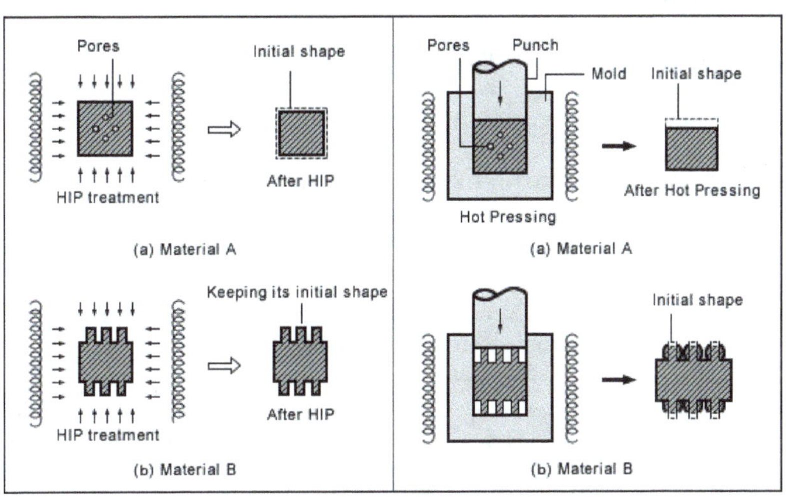

Compared to hot pressing, HIP can provide material shapes not much different from the initial one after pressure. A material even after changing its shape can keep its initial shape, and will be relatively less restricted by processing of products. By making full use of these features, HIP has been applied in various fields.

Pressure Medium Gas (Argon Gas) under High Pressure

Argon gas at 1000°C and under pressure of 98MPa is likely to cause intense convection due to low density and coefficients of viscosity (30% and 15% of water, respectively), and high coefficients of thermal expansion. Therefore, heat transfer coefficients of HIP equipment become higher than that of an ordinary electronic furnace.

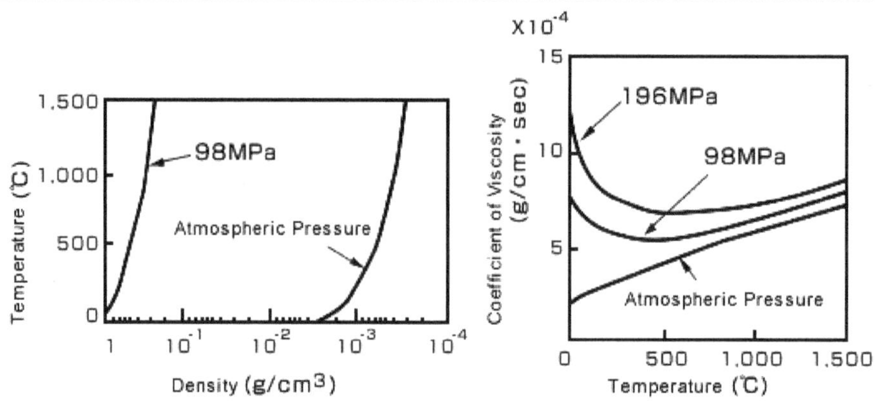

(a) Density of Argon Gas (b) Viscosity of Argon Gas

HIP Application

HIP is applied in a wide range of fields as follows:

- Pressure sintering of powder.
- Diffusion bonding of different types of materials.
- Removal of residual pores in sintered items.
- Removal of inner defects of castings.
- Rejuvenation of parts damaged by fatigue or creep.
- High pressure impregnated carbonization method.

	Applied Technology	Practical Application	Material under Study
1	Pressure sintering of powder.	PM high speed steel (jigs), Ni-based superalloys (engine turbine discs), Ti alloys (aircraft components), Cr (target).	Ti-Ni alloy, amorphous metal, Si_3N_4, SiC.
2	Diffusion bonding(Production of composites).	Nuclear fuel assemblies (nuclear reactors), B fiber-Al alloy composites (space shuttle struts), various corrosion/abrasion-resistant alloy composite parts (valves for corrosive gas, mill roll, cylinders for injection molding machines etc.)	SiC-Al alloy composites, Nb_3Sn-copper, Si_3N_4-stainless steel.

3	Removal of residual pores in sintered items.	Cemented carbide (jigs), Al_2o_3(cutting tools), Al_2o_3-TiC (cutting tools), soft ferrites (magnetic head), Si_3N_4 (bearings and ceramics structure).	SiC, PSZ, ZnSe.
4	Removal of inner defects of castings.	Al alloys, Ni-base superalloys (jet engine turbine blades), Ti alloys (aircraft components), 17-4PH stainless steel.	
5	Rejuvenation of parts damaged by fatigue or creep.	Superalloy precision castings (gas turbine blades).	
6	High pressure impregnated carbonization method.	Carbon composites.	

HIP Treatment

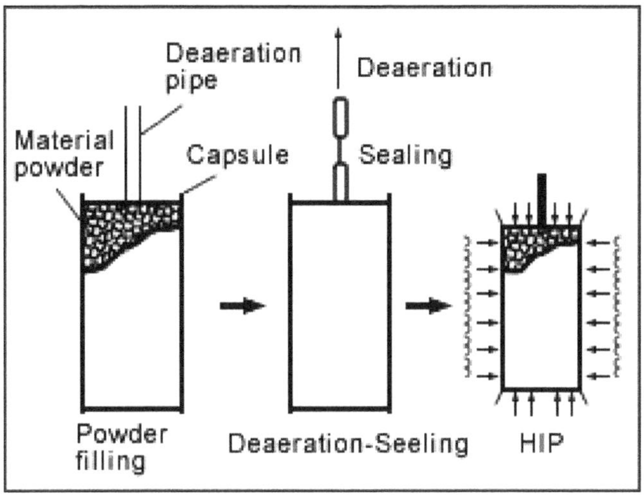

Materials need various treatment depending on the situations. The most typical methods include 'Capsule Method' and 'Capsule Free Method'.

'Capsule Method', is to carry out HIP after enclosing powder or a body molded from powder in a gastight capsule and evacuating the capsule.

This 'Capsule Method' can provide high densification even for materials that are difficult to sinter by ordinary sintering techniques. Therefore, it is most commonly adopted as pressure sintering process of powder materials. It's also used for diffusion bonding of different types of materials or high pressure impregnated carbonization methods.

The following Table gives a summary of main materials for capsule free methods and HIP treatment temperature/pressure.

Material	Temperature	Pressure
Powder high-speed steel	1,000 to 1,200°C	to 100MPa
Ni base alloy	1,170 to 1,280°C	100 to 150MPa
Ti alloy (Ti-6Al-4V)	880 to 960°C	to 100MPa
Cr	1,200 to 1,300°C	to 100MPa
Cu Alloy	500 to 900°C	to 100MPa

Al Alloy	350 to 500°C	to 100MPa
Cemented carbide (WC-Co)	1,300 to 1,350°C	30 to 100MPa
Ti Ba O_3	1,000 to 1,200°C	to 100MPa
PZT	950 to 1,150°C	to 100MPa
Ni-Zn-ferrite	1,050 to 1,180°C	to 100MPa
Mn-Zn-ferrite	1,180 to 1,250°C	to 100MPa
Al_2O_3	1,350 to 1,450°C	to 100MPa
Y-PSZ (yttria partially-stabilized zirconia)	1,350 to 1,500°C	to 100MPa
Si_3N_4-Al_2O_3-Y_2O_3	1,700 to 1,800°C	to 100MPa
SiC	1,950 to 1,050°C	100 to 200MPa

Meanwhile, if the pores inside the material are isolated, closed off, and not connected to the surface of the material, these pores might be squashed and eliminated by HIP treatment. On the other hand, open pores that connect with the surface of the material will not be squashed even after HIP treatment. Therefore, HIP treatment to the material with the closed pores can provide high densification of the entire material.

Such materials require no capsules for HIP treatment, which is called 'Capsule Free Method'. This is used for removal of residual pores on sintered items, removal of inner defects of castings, and rejuvenation of parts damaged by fatigue or creep.

References

- "Application Overview: Injection Molding". Yaskawa America, Inc. Archived from the original on 2006-04-12. Retrieved 2009-02-27
- Moulding-process, engineering, topics: sciencedirect.com, Retrieved 16 March, 2019
- Jan Schroers; Thomas M. Hodges; Golden Kumar; Hari Raman; Anthony J. Barnes; Quoc Pham; Theodore A. Waniuk (February 2011). "Thermoplastic blow molding of metals". Materials Today. 14: 14–19. Doi:10.1016/S1369-7021(11)70018-9
- What-is-dip-molding: monroeengineering.com, Retrieved 17 April, 2019
- Pascault, Jean-Pierre; Sautereau, Henry; Verdu, Jacques; Williams, Roberto J. J. (2002-02-20). Thermosetting Polymers. CRC Press. ISBN 9780824744052.
- What-is-laminating, jessup-manufacturing-news-and-events: jessupmfg.com, Retrieved 18 May, 2019
- Kang, Moon Koo; Lee, Woo Il; Hahn, H. Thomas (2000-09-01). "Formation of microvoids during resin-transfer molding process". Composites Science and Technology. 60 (12–13): 2427–2434. Doi:10.1016/S0266-3538(00)00036-1

Forming Processes

CHAPTER 5

Forming is a manufacturing process where the shape of a partly finished product is changed using plastic deformation. It includes processes such as tube beading, forging, stamping, shearing, hydroforming, decambering, hemming, explosive forming, etc. This chapter discusses in detail these processes related to forming.

Forming, also known as "metal forming," includes a wide range of manufacturing processes in which metal is deformed into a required shape by the application of suitable stresses. To make the metal plastically deformed, forces must be applied that are greater than the yield strength of the metals. The magnitude of the compression, stretching or bending in the material, is directly proportional to the force applied.

In industrial processes, forming is characterized by:

- High levels of loads and stresses ranging from 50 to 2,500 newtons per square millimeter.
- Big, heavy and expensive machinery, which can generate a high amount of stresses and loads.
- Many parts that are produced in less time, which helps in maximizing the production economy or reaching economies of scale.

TUBE BEADING

Tube beading is the process of crimping to the end of a cylindrical metal tube to form a spout shape. Beaded tube ends are commonly used to connect hoses or slide into other metal tubes. Some tube beading is done simply to strengthen and protect the shape of the rolled metal form. Metalworkers use internal roll forming or ram forming to bend the ends of metal tubes into the bead shape. Many different lengths and diameters of beaded metal tubes are used in the construction, manufacturing and automotive industries.

Manufacturers commonly connect beaded tube ends to create rigid metal forms. Matched tubes are sized to fit inside one another. Connecting tubes are slid together at their beaded ends to form a snug joint. Pressure from the raised lip at the end of the tube holds the joint together. Bolts and welds are often used to fasten beaded tube ends for permanent assembly.

Internal roll forming raises the beaded section by applying pressure from the inside while the tube is held in place. Manual and mechanical devices are used to roll around the diameter of the tube and create the bead in a relatively slow process. The internal roll forming process is typically used for small scale beading and custom tube beading. This method of tube beading is generally considered to be more precise. Some heavy gauge metal tubes will not fit into manual internal roll forming tools.

Ram forming is a faster tube beading procedure that presses internally and externally with a set of punches. Tube ends are formed in a single motion as they are pressed onto the hard metal punches. Large-scale tube beading is done almost exclusively using the ram forming process. The quick ramming process flexes the end of the tube causing the metal to heat up. Beads formed by ramming punches may be slightly uneven or cause minor structural variations in the tube ends.

Either style of tube beading forms a nipple shape at the end of a section of metal tubing that is ideal for connecting to a flexible hose. Hoses attached to beaded ends are often clamped in place. Adjustable screw clamps are commonly used to create an air and watertight seal between the hose and the tube. Adhesives and polymer sealants applied to the beaded ends provide increased protection against leakage. If the tube end becomes damaged, it can usually be bent back into shape to prevent hose splitting.

FORGING

Forging is a manufacturing process involving the shaping of metal using localized compressive forces. The blows are delivered with a hammer (often a power hammer) or a die. Forging is often classified according to the temperature at which it is performed: cold forging (a type of cold working), warm forging, or hot forging (a type of hot working). For the latter two, the metal is heated, usually in a forge. Forged parts can range in weight from less than a kilogram to hundreds of metric tons. Forging has been done by smiths for millennia; the traditional products were kitchenware, hardware, hand tools, edged weapons, cymbals, and jewellery. Since the Industrial Revolution, forged parts are widely used in mechanisms and machines wherever a component requires high strength; such forgings usually require further processing (such as machining) to achieve a finished part. Today, forging is a major worldwide industry.

Hot metal ingot being loaded into a hammer forge.

Advantages and Disadvantages

Forging can produce a piece that is stronger than an equivalent cast or machined part. As the metal is shaped during the forging process, its internal grain texture deforms to follow the general shape of the part. As a result, the texture variation is continuous throughout the part, giving rise to a piece with improved strength characteristics. Additionally, forgings can achieve a lower total cost than casting or fabrication. Considering all the costs that are incurred in a product's life cycle from procurement to lead time to rework, and factoring in the costs of scrap, and downtime and other quality considerations, the long-term benefits of forgings can outweigh the short-term cost savings that castings or fabrications might offer.

Some metals may be forged cold, but iron and steel are almost always hot forged. Hot forging prevents the work hardening that would result from cold forming, which would increase the difficulty of performing secondary machining operations on the piece. Also, while work hardening may be desirable in some circumstances, other methods of hardening the piece, such as heat treating, are generally more economical and more controllable. Alloys that are amenable to precipitation hardening, such as most aluminium alloys and titanium, can be hot forged, followed by hardening.

Production forging involves significant capital expenditure for machinery, tooling, facilities and personnel. In the case of hot forging, a high-temperature furnace (sometimes referred to as the forge) is required to heat ingots or billets. Owing to the size of the massive forging hammers and presses and the parts they can produce, as well as the dangers inherent in working with hot metal, a special building is frequently required to house the operation. In the case of drop forging operations, provisions must be made to absorb the shock and vibration generated by the hammer. Most forging operations use metal-forming dies, which must be precisely machined and carefully heat-treated to correctly shape the workpiece, as well as to withstand the tremendous forces involved.

Processes

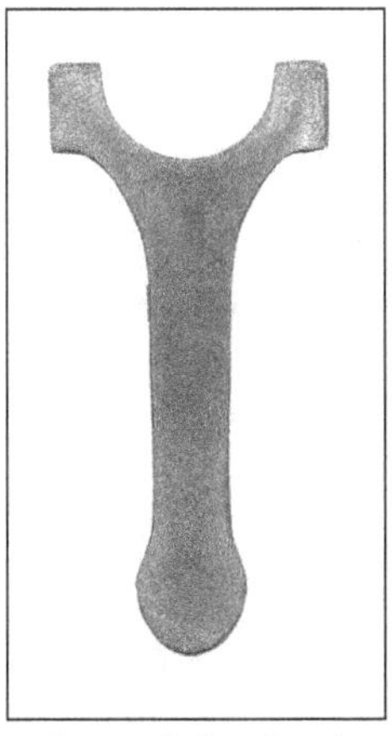

A cross-section of a forged connecting rod that has been etched to show the grain flow.

There are many different kinds of forging processes available; however, they can be grouped into three main classes:

- Drawn out: Length increases, cross-section decreases.

- Upset: Length decreases, cross-section increases.
- Squeezed in closed compression dies: produces multidirectional flow.

Common forging processes include: roll forging, swaging, cogging, open-die forging, impression-die forging, press forging, automatic hot forging and upsetting.

Temperature

All of the following forging processes can be performed at various temperatures; however, they are generally classified by whether the metal temperature is above or below the recrystallization temperature. If the temperature is above the material's recrystallization temperature it is deemed *hot forging*; if the temperature is below the material's recrystallization temperature but above 30% of the recrystallization temperature (on an absolute scale) it is deemed *warm forging*; if below 30% of the recrystallization temperature (usually room temperature) then it is deemed *cold forging*. The main advantage of hot forging is that it can be done more quickly and precisely, and as the metal is deformed work hardening effects are negated by the recrystallization process. Cold forging typically results in work hardening of the piece.

Drop Forging

Drop forging is a forging process where a hammer is raised and then "dropped" onto the workpiece to deform it according to the shape of the die. There are two types of drop forging: open-die drop forging and closed-die drop forging. As the names imply, the difference is in the shape of the die, with the former not fully enclosing the workpiece, while the latter does.

Open-die Drop Forging

Open-die drop forging (with two dies) of an ingot to be further processed into a wheel.

Open-die forging is also known as *smith forging*. In open-die forging, a hammer strikes and deforms the workpiece, which is placed on a stationary anvil. Open-die forging gets its name from the fact that the dies (the surfaces that are in contact with the workpiece) do not enclose the workpiece, allowing it to flow except where contacted by the dies. The operator therefore needs to orient and position the workpiece to get the desired shape. The dies are usually flat in shape, but some have a specially shaped surface for specialized operations. For example, a die may have a round, concave, or convex surface or be a tool to form holes or be a cut-off tool. Open-die forgings can be worked into shapes which include discs, hubs, blocks, shafts (including step shafts or with flanges), sleeves, cylinders, flats, hexes, rounds, plate, and some custom shapes. Open-die forging lends itself to short runs and is appropriate for art smithing and custom work. In some cases, open-die forging may be employed to rough-shape ingots to prepare them for subsequent operations. Open-die forging may also orient the grain to increase strength in the required direction.

Advantages of open-die Forging

- Reduced chance of voids.
- Better fatigue resistance.
- Improved microstructure.
- Continuous grain flow.
- Finer grain size.
- Greater strength.

"Cogging" is the successive deformation of a bar along its length using an open-die drop forge. It is commonly used to work a piece of raw material to the proper thickness. Once the proper thickness is achieved the proper width is achieved via "edging". "Edging" is the process of concentrating material using a concave shaped open-die. The process is called "edging" because it is usually carried out on the ends of the workpiece. "Fullering" is a similar process that thins out sections of the forging using a convex shaped die. These processes prepare the workpieces for further forging processes.

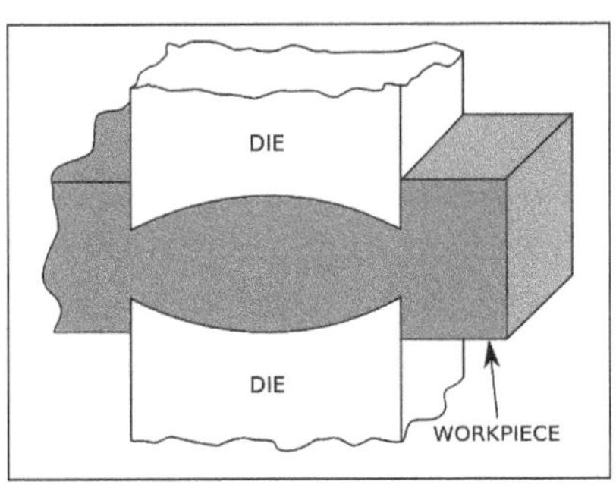

Edging.

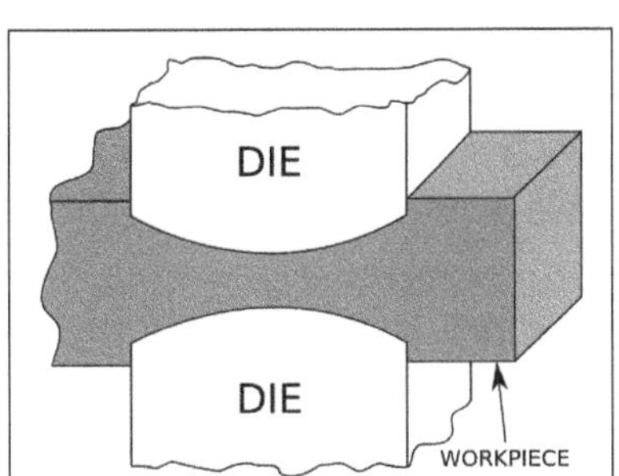

Fullering.

Impression-die Forging

Impression-die forging is also called "closed-die forging". In impression-die forging, the metal is placed in a die resembling a mold, which is attached to an anvil. Usually, the hammer die is shaped as well. The hammer is then dropped on the workpiece, causing the metal to flow and fill the die cavities. The hammer is generally in contact with the workpiece on the scale of milliseconds. Depending on the size and complexity of the part, the hammer may be dropped multiple times in quick succession. Excess metal is squeezed out of the die cavities, forming what is referred to as "flash". The flash cools more rapidly than the rest of the material; this cool metal is stronger than the metal in the die, so it helps prevent more flash from forming. This also forces the metal to completely fill the die cavity. After forging, the flash is removed.

In commercial impression-die forging, the workpiece is usually moved through a series of cavities in a die to get from an ingot to the final form. The first impression is used to distribute the metal into the rough shape in accordance to the needs of later cavities; this impression is called an "edging", "fullering", or "bending" impression. The following cavities are called "blocking" cavities, in which the piece is working into a shape that more closely resembles the final product. These stages usually impart the workpiece with generous bends and large fillets. The final shape is forged in a "final" or "finisher" impression cavity. If there is only a short run of parts to be done, then it may be more economical for the die to lack a final impression cavity and instead machine the final features.

Impression-die forging has been improved in recent years through increased automation which includes induction heating, mechanical feeding, positioning and manipulation, and the direct heat treatment of parts after forging. One variation of impression-die forging is called "flashless forging", or "true closed-die forging". In this type of forging, the die cavities are completely closed, which keeps the workpiece from forming flash. The major advantage to this process is that less metal is lost to flash. Flash can account for 20 to 45% of the starting material. The disadvantages of this process include additional cost due to a more complex die design and the need for better lubrication and workpiece placement.

There are other variations of part formation that integrate impression-die forging. One method incorporates casting a forging preform from liquid metal. The casting is removed after it has solidified, but while still hot. It is then finished in a single cavity die. The flash is trimmed, then the part is quench hardened. Another variation follows the same process as outlined above, except the preform is produced by the spraying deposition of metal droplets into shaped collectors (similar to the Osprey process).

Closed-die forging has a high initial cost due to the creation of dies and required design work to make working die cavities. However, it has low recurring costs for each part, thus forgings become more economical with greater production volume. This is one of the major reasons closed-die forgings are often used in the automotive and tool industries. Another reason forgings are common in these industrial sectors is that forgings generally have about a 20 percent higher strength-to-weight ratio compared to cast or machined parts of the same material.

Design of Impression-die Forgings and Tooling

Forging dies are usually made of high-alloy or tool steel. Dies must be impact- and wear-resistant, maintain strength at high temperatures, have the ability to withstand cycles of rapid heating and cooling. In order to produce a better, more economical die the following standards are maintained:

- The dies part along a single, flat plane whenever possible. If not, the parting plane follows the contour of the part.
- The parting surface is a plane through the center of the forging and not near an upper or lower edge.
- Adequate draft is provided; usually at least 3° for aluminium and 5° to 7° for steel.
- Generous fillets and radii are used.
- Ribs are low and wide.
- The various sections are balanced to avoid extreme difference in metal flow.
- Full advantage is taken of fiber flow lines.
- Dimensional tolerances are not closer than necessary.

The dimensional tolerances of a steel part produced using the impression-die forging method are outlined in the table below. The dimensions across the parting plane are affected by the closure of the dies, and are therefore dependent on die wear and the thickness of the final flash. Dimensions that are completely contained within a single die segment or half can be maintained at a significantly greater level of accuracy.

Dimensional Tolerances for Impression-die Forgings

Mass [kg (lb)]	Minus tolerance [mm (in)]	Plus tolerance [mm (in)]
0.45 (1)	0.15 (0.006)	0.46 (0.018)
0.91 (2)	0.20 (0.008)	0.61 (0.024)
2.27 (5)	0.25 (0.010)	0.76 (0.030)
4.54 (10)	0.28 (0.011)	0.84 (0.033)
9.07 (20)	0.33 (0.013)	0.99 (0.039)
22.68 (50)	0.48 (0.019)	1.45 (0.057)
45.36 (100)	0.74 (0.029)	2.21 (0.087)

A lubricant is used when forging to reduce friction and wear. It is also used as a thermal barrier to restrict heat transfer from the workpiece to the die. Finally, the lubricant acts as a parting compound to prevent the part from sticking in the dies.

Press Forging

Press forging works by slowly applying a continuous pressure or force, which differs from the near-instantaneous impact of drop-hammer forging. The amount of time the dies are in contact with the workpiece is measured in seconds (as compared to the milliseconds of drop-hammer forges). The press forging operation can be done either cold or hot.

The main advantage of press forging, as compared to drop-hammer forging, is its ability to deform the complete workpiece. Drop-hammer forging usually only deforms the surfaces of the work piece in contact with the hammer and anvil; the interior of the workpiece will stay relatively undeformed. Another advantage to the process includes the knowledge of the new part's strain rate. By controlling the compression rate of the press forging operation, the internal strain can be controlled.

There are a few disadvantages to this process, most stemming from the workpiece being in contact with the dies for such an extended period of time. The operation is a time-consuming process due to the amount and length of steps. The workpiece will cool faster because the dies are in contact with workpiece; the dies facilitate drastically more heat transfer than the surrounding atmosphere. As the workpiece cools it becomes stronger and less ductile, which may induce cracking if deformation continues. Therefore, heated dies are usually used to reduce heat loss, promote surface flow, and enable the production of finer details and closer tolerances. The workpiece may also need to be reheated.

When done in high productivity, press forging is more economical than hammer forging. The operation also creates closer tolerances. In hammer forging a lot of the work is absorbed by the machinery; when in press forging, the greater percentage of work is used in the work piece. Another advantage is that the operation can be used to create any size part because there is no limit to the size of the press forging machine. New press forging techniques have been able to create a higher degree of mechanical and orientation integrity. By the constraint of oxidation to the outer layers of the part, reduced levels of microcracking occur in the finished part.

Press forging can be used to perform all types of forging, including open-die and impression-die forging. Impression-die press forging usually requires less draft than drop forging and has better dimensional accuracy. Also, press forgings can often be done in one closing of the dies, allowing for easy automation.

Upset Forging

Upset forging increases the diameter of the workpiece by compressing its length. Based on number of pieces produced, this is the most widely used forging process. A few examples of common parts produced using the upset forging process are engine valves, couplings, bolts, screws, and other fasteners.

Upset forging is usually done in special high-speed machines called *crank presses*. The machines are usually set up to work in the horizontal plane, to facilitate the quick exchange of workpieces from one station to the next, but upsetting can also be done in a vertical crank press or a hydraulic press. The initial workpiece is usually wire or rod, but some machines can accept bars up to 25 cm (9.8 in) in diameter and a capacity of over 1000 tons. The standard upsetting machine employs split dies that contain multiple cavities. The dies open enough to allow the workpiece to move from one cavity to the next; the dies then close and the heading tool, or ram, then moves longitudinally against the bar, upsetting it into the cavity. If all of the cavities are utilized on every cycle, then a finished part will be produced with every cycle, which makes this process advantageous for mass production.

These rules must be followed when designing parts to be upset forged:

- The length of unsupported metal that can be upset in one blow without injurious buckling should be limited to three times the diameter of the bar.

- Lengths of stock greater than three times the diameter may be upset successfully, provided that the diameter of the upset is not more than 1.5 times the diameter of the stock.
- In an upset requiring stock length greater than three times the diameter of the stock, and where the diameter of the cavity is not more than 1.5 times the diameter of the stock, the length of unsupported metal beyond the face of the die must not exceed the diameter of the bar.

Automatic Hot Forging

The automatic hot forging process involves feeding mill-length steel bars (typically 7 m (23 ft) long) into one end of the machine at room temperature and hot forged products emerge from the other end. This all occurs rapidly; small parts can be made at a rate of 180 parts per minute (ppm) and larger can be made at a rate of 90 ppm. The parts can be solid or hollow, round or symmetrical, up to 6 kg (13 lb), and up to 18 cm (7.1 in) in diameter. The main advantages to this process are its high output rate and ability to accept low-cost materials. Little labor is required to operate the machinery.

There is no flash produced so material savings are between 20 and 30% over conventional forging. The final product is a consistent 1,050 °C (1,920 °F) so air cooling will result in a part that is still easily machinable (the advantage being the lack of annealing required after forging). Tolerances are usually ±0.3 mm (0.012 in), surfaces are clean, and draft angles are 0.5 to 1°. Tool life is nearly double that of conventional forging because contact times are on the order of 0.06-second. The downside is that this process is only feasible on smaller symmetric parts and cost; the initial investment can be over $10 million, so large quantities are required to justify this process.

The process starts by heating the bar to 1,200 to 1,300 °C (2,190 to 2,370 °F) in less than 60 seconds using high-power induction coils. It is then descaled with rollers, sheared into blanks, and transferred through several successive forming stages, during which it is upset, preformed, final forged, and pierced (if necessary). This process can also be coupled with high-speed cold-forming operations. Generally, the cold forming operation will do the finishing stage so that the advantages of cold-working can be obtained, while maintaining the high speed of automatic hot forging.

Examples of parts made by this process are: wheel hub unit bearings, transmission gears, tapered roller bearing races, stainless steel coupling flanges, and neck rings for LP gas cylinders. Manual transmission gears are an example of automatic hot forging used in conjunction with cold working.

Roll Forging

Roll forging is a process where round or flat bar stock is reduced in thickness and increased in length. Roll forging is performed using two cylindrical or semi-cylindrical rolls, each containing one or more shaped grooves. A heated bar is inserted into the rolls and when it hits a spot the rolls rotate and the bar is progressively shaped as it is rolled through the machine. The piece is then transferred to the next set of grooves or turned around and reinserted into the same grooves. This continues until the desired shape and size is achieved. The advantage of this process is there is no flash and it imparts a favorable grain structure into the workpiece.

Examples of products produced using this method include axles, tapered levers and leaf springs.

Net-shape and Near-net-shape Forging

This process is also known as *precision forging*. It was developed to minimize cost and waste associated with post-forging operations. Therefore, the final product from a precision forging needs little or no final machining. Cost savings are gained from the use of less material, and thus less scrap, the overall decrease in energy used, and the reduction or elimination of machining. Precision forging also requires less of a draft, 1° to 0°. The downside of this process is its cost, therefore it is only implemented if significant cost reduction can be achieved.

Cold Forging

Near net shape forging is most common when parts are forged without heating the slug, bar or billet. Aluminum is a common material that can be cold forged depending on final shape. Lubrication of the parts being formed is critical to increase the life of the mating dies.

Cost Implications

To achieve a low-cost net shape forging for demanding applications that are subject to a high degree of scrutiny, i.e. non-destructive testing by way of a dye-penetrant inspection technique, it is crucial that basic forging process disciplines be implemented. If the basic disciplines are not met, subsequent material removal operations will likely be necessary to remove material defects found at non-destructive testing inspection. Hence low-cost parts will not be achievable.

Example disciplines are: die-lubricant management (Use of uncontaminated and homogeneous mixtures, amount and placement of lubricant). Tight control of die temperatures and surface finish / friction.

Induction Forging

Unlike the above processes, induction forging is based on the type of heating style used. Many of the above processes can be used in conjunction with this heating method.

Multidirectional Forging

Multidirectional Forging is forming of a work piece in a single step in several directions. The multidirectional forming takes place through constructive measures of the tool. The vertical movement of the press ram is redirected using wedges which distributes and redirects the force of the forging press in horizontal directions.

Isothermal Forging

Isothermal forging is a process by which the materials and the die are heated to the same temperature (*iso-* meaning "equal"). Adiabatic heating is used to assist in the deformation of the material, meaning the strain rates are highly controlled. Commonly used for forging aluminum, which has a lower forging temperature than steels. Forging temperatures for Aluminum are around 800 °F, while steels and super alloys can be 1700-2300 °F.

Benefits:

- Near net shapes which lead to lower machining requirements and therefore lower scrap rates.
- Reproducibility of the part.
- Due to the lower heat loss smaller machines can be used to make the forging.

Disadvantages:

- Higher die material costs to handle temperatures and pressures.
- Uniform heating systems are required.
- Protective atmospheres or vacuum to reduce oxidation of the dies and material.
- Low production rates.

Materials and Applications

Forging of Steel

Depending on the forming temperature steel forging can be divided into:

- Hot forging of steel.
    - Forging temperatures above the recrystallization temperature between 950–1250 °C.
    - Good formability.
    - Low forming forces.
    - Constant tensile strength of the workpieces.
- Warm forging of steel.
    - Forging temperatures between 750–950 °C.
    - Less or no scaling at the workpiece surface.
    - Narrower tolerances achievable than in hot forging.
    - Limited formability and higher forming forces than for hot forging.
    - Lower forming forces than in cold forming.
- Cold forging of steel.
    - Forging temperatures at room conditions, self-heating up to 150 °C due to the forming energy.
    - Narrowest tolerances achievable.

- No scaling at workpiece surface.
- Increase of strength and decrease of ductility due to strain hardening.
- Low formability and high forming forces are necessary.

For industrial processes steel alloys are primarily forged in hot condition. Brass, bronze, copper, precious metals and their alloys are manufactured by cold forging processes, while each metal requires a different forging temperature.

Forging of Aluminium

- Aluminium forging is performed at a temperature range between 350–550 °C.
- Forging temperatures above 550 °C are too close to the solidus temperature of the alloys and lead in conjunction with varying effective strains to unfavorable workpiece surfaces and potentially to a partial melting as well as fold formation.
- Forging temperatures below 350 °C reduce formability by increasing the yield stress, which can lead to unfilled dies, cracking at the workpiece surface and increased die forces.

Due to the narrow temperature range and high thermal conductivity, aluminium forging can only be realized in a particular process window. To provide good forming conditions a homogeneous temperature distribution in the entire workpiece is necessary. Therefore, the control of the tool temperature has a major influence to the process. For example, by optimizing the preform geometries the local effective strains can be influenced to reduce local overheating for a more homogeneous temperature distribution.

Application of Aluminium Forged Parts

High-strength aluminium alloys have the tensile strength of medium strong steel alloys while providing significant weight advantages. Therefore, aluminium forged parts are mainly used in aerospace, automotive industry and many other fields of engineering especially in those fields, where highest safety standards against failure by abuse, by shock or vibratory stresses are needed. Such parts are for example pistons, chassis parts, steering components and brake parts. Commonly used alloys are AlSi1MgMn (EN AW-6082) and AlZnMgCu1,5 (EN AW-7075). About 80% of all aluminium forged parts are made of AlSi1MgMn. The high-strength alloy AlZnMgCu1,5 is mainly used for aerospace applications.

Equipment

The most common type of forging equipment is the hammer and anvil. Principles behind the hammer and anvil are still used today in *drop-hammer* equipment. The principle behind the machine is simple: raise the hammer and drop it or propel it into the workpiece, which rests on the anvil. The main variations between drop-hammers are in the way the hammer is powered; the most common being air and steam hammers. Drop-hammers usually operate in a vertical position. The main reason for this is excess energy (energy that isn't used to deform the workpiece) that isn't released as heat or sound needs to be transmitted to the foundation. Moreover, a large machine base is needed to absorb the impacts.

Hydraulic drop-hammer.

To overcome some shortcomings of the drop-hammer, the *counterblow machine* or *impactor* is used. In a counterblow machine both the hammer and anvil move and the workpiece is held between them. Here excess energy becomes recoil. This allows the machine to work horizontally and have a smaller base. Other advantages include less noise, heat and vibration. It also produces a distinctly different flow pattern. Both of these machines can be used for open-die or closed-die forging.

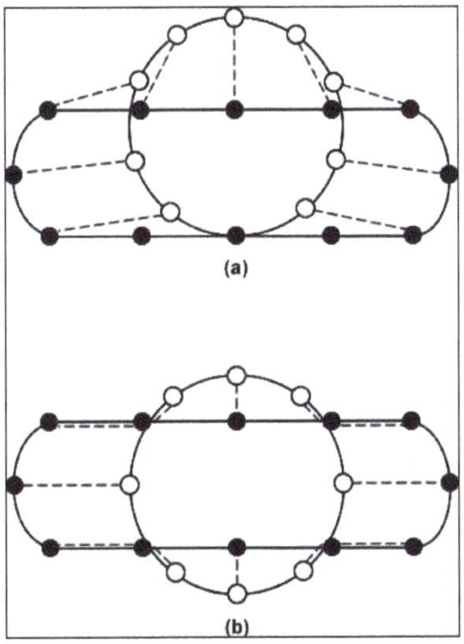

(a) Material flow of a conventionally forged disc; (b) Material flow of an impactor forged disc.

Forging Presses

A forging press, often just called a press, is used for press forging. There are two main types: mechanical and hydraulic presses. Mechanical presses function by using cams, cranks and/or toggles to produce a preset (a predetermined force at a certain location in the stroke) and reproducible stroke. Due to the nature of this type of system, different forces are available at different stroke positions. Mechanical presses are faster than their hydraulic counterparts (up to 50 strokes per minute). Their capacities range from 3 to 160 MN (300 to 18,000 short tons-force). Hydraulic presses use fluid pressure and a piston to generate force. The advantages of a hydraulic press over a mechanical press are its flexibility and greater capacity. The disadvantages include a slower, larger, and costlier machine to operate.

The roll forging, upsetting, and automatic hot forging processes all use specialized machinery.

List of Large Forging Presses, by Ingot Size

Force (tonnes)	Ingot size (tonnes)	Company	Location
16,500	600	Shanghai Electric Group	Shanghai, China
16,000	600	China National Erzhong Group	Deyang, China
14,000	600	Japan Steel Works	Japan
15,000	580	China First Heavy Industries Group	Heilongjiang, China
13,000		Doosan	South Korea

List of Large Forging Presses, by Force

Force (tonnes)	Force (tons)	Ingot size (tonnes)	Company	Location
80,000	(88,200)	>150	China Erzhong	Deyang, China
75,000	(82,690)		VSMPO-AVISMA	Russia
65,000	(71,660)		Aubert & Duval	Issoire, France
53,500	(60,000)		Weber Metals, Inc.	California, United States
(45,350)	50,000	20	Alcoa, Wyman Gordon	USA
40,000	(44,100)		Aubert & Duval	Pamiers, France
30,000	(33,080)	8	Wyman Gordon	Livingston, Scotland
30,000	(33,070)		Weber Metals, Inc.	California, United States
30,000	(30,108)		Firth Rixson	Georgia, United States

SHEARING

Shearing, also known as die cutting, is a process which cuts stock without the formation of chips or the use of burning or melting. Strictly speaking, if the cutting blades are straight the process is called shearing; if the cutting blades are curved then they are shearing-type operations. The most commonly sheared materials are in the form of sheet metal or plates, however rods can also be sheared. Shearing-type operations include: blanking, piercing, roll slitting, and trimming. It is used in metalworking and also with paper and plastics.

Principle

A punch (or moving blade) is used to push a workpiece against the die (or fixed blade), which is fixed. Usually the clearance between the two is 5 to 40% of the thickness of the material, but dependent on the material. Clearance is defined as the separation between the blades, measured at the point where the cutting action takes place and perpendicular to the direction of blade movement. It affects the finish of the cut (burr) and the machine's power consumption. This causes the material to experience highly localized shear stresses between the punch and die. The material will then fail when the punch has moved 15 to 60% the thickness of the material, because the shear stresses are greater than the shear strength of the material and the remainder of the material is torn. Two distinct sections can be seen on a sheared workpiece, the first part being plastic deformation and the second being fractured. Because of normal inhomogeneities in materials and inconsistencies in clearance between the punch and die, the shearing action does not occur in a uniform manner. The fracture will begin at the weakest point and progress to the next weakest point until the entire workpiece has been sheared; this is what causes the rough edge. The rough edge can be reduced if the workpiece is clamped from the top with a die cushion. Above a certain pressure the fracture zone can be completely eliminated. However, the sheared edge of the workpiece will usually experience workhardening and cracking. If the workpiece has too much clearance, then it may experience roll-over or heavy burring.

Tool Materials

- Low alloy steel is used in low production of materials that range up to 0.64 cm (1/4 in) thick.
- High-carbon, high chromium steel is used in high production of materials that also range up to 0.64 cm (1/4 in) in thickness.
- Shock-resistant steel is used in materials that are equal to 0.64 cm (1/4 in) thick or more.

Tolerances and Surface Finish

When shearing a sheet, the typical tolerance is +0.1 inches or -0.1 inches, but it is feasible to get the tolerance to within +0.005 inches or -0.005 inches. While shearing a bar and angle, the typical tolerance is +0.06 inches or -0.06 inches, but it is possible to get the tolerance to +0.03 inches or -0.03 inches. Surface finishes typically occur within the 250 to 1000 microinches range, but can range from 125 to 2000 microinches. A secondary operation is required if one wants better surfaces than this.

STAMPING

Metal stamping is a manufacturing process used to convert flat metal sheets into specific shapes. It is a complex process that can include a number of metal forming techniques — blanking, punching, bending and piercing, to name a few.

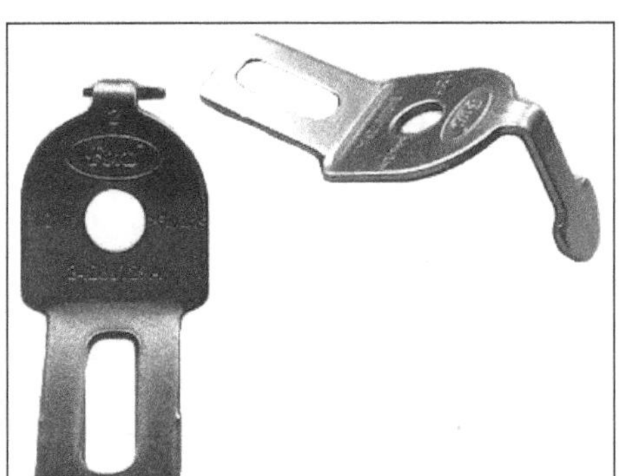

Metal stamping is a fast and cost-effective solution for this large-quantity manufacturing need.

Manufacturers who need metal parts stamped for a project generally look for three important qualities:

- High quality/durability.
- Low Cost.
- Fast turnaround time.

Stamping Basics

Stamping — also called pressing — involves placing flat sheet metal, in either coil or blank form, into a stamping press. In the press, a tool and die surface form the metal into the desired shape. Punching, blanking, bending, coining, embossing, and flanging are all stamping techniques used to shape the metal.

Before the material can be formed, stamping professionals must design the tooling via CAD/CAM engineering technology. These designs must be as precise as possible to ensure each punch and bend maintains proper clearance and, therefore, optimal part quality. A single tool 3D model can contain hundreds of parts, so the design process is often quite complex and time-consuming.

Once the tool's design is established, a manufacturer can use a variety of machining, grinding, wire EDM and other manufacturing services to complete its production.

Types of Metal Stamping

There are three major types of metal stamping techniques: progressive, fourslide and deep draw.

Progressive Die Stamping

Progressive die stamping features a number of stations, each with a unique function.

First, strip metal is feds through a progressive stamping press. The strip unrolls steadily from a coil and into the die press, where each station in the tool then performs a different cut, punch, or bend. The actions of each successive station add onto the work of the previous stations, resulting in a completed part.

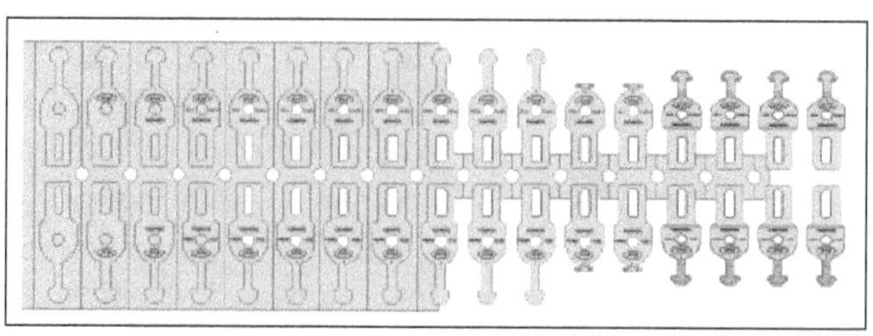

A manufacturer might have to repeatedly change the tool on a single press or occupy a number of presses, each performing one action required for a completed part. Even using multiple presses, secondary machining services were often required to truly complete a part. For that reason, progressive die stamping is the ideal solution for metal parts with complex geometry to meet:

- Faster turnaround.
- Lower labor cost.
- Shorter run length.
- Higher repeatability.

Fourslide Stamping

Fourslide, or multi-slide, involves horizontal alignment and four different slides; in other words, four tools are used simultaneously to shape the workpiece. This process allows for intricate cuts and complex bends to develop even the most complex parts.

Fourslide metal stamping can offer several advantages over traditional press stamping that make it an ideal choice for many applications. Some of these advantages include:

- Versatility for more complex parts.
- More flexibility for design changes.

As its name implies, a fourslide has four slides — meaning that up to four different tools, one per slide, can be used to achieve multiple bends simultaneously. As material feeds into a fourslide, it is bent in quick succession by each shaft that is equipped with a tool.

Deep Draw Stamping

Deep drawing involves pulling a sheet metal blank into the die via a punch, forming it into a shape. The method is referred to as "deep drawing" when the depth of the drawn part exceeds its diameter. This type of forming is ideal for creating components that need several series of diameters and is a cost-effective alternative to turning processes, which typically require using up more raw materials. Common applications and products made from deep drawing include:

- Automotive components.
- Aircraft parts.
- Electronic relays.
- Utensils and cookware.

Short Run Stamping

Short run metal stamping requires minimal upfront tooling expenses and can be an ideal solution for prototypes or small projects. After the blank is created, manufacturers use a combination of custom tooling components and die inserts to bend, punch or drill the part. The custom forming operations and smaller run size can result in a higher per-piece charge, but the absence of tooling costs can make short run more cost-efficient for many projects, especially those requiring fast turnaround.

Manufacturing Tools for Stamping

There are several steps in producing metal stamping. The first step is designing and manufacturing the actual tool used to create the product.

Stock Strip Layout & Design: A designer uses is used to design the strip and determine dimensions, tolerances, feed direction, scrap minimization and more.

Tool Steel and Die Set Machining: CNC ensures a higher level of precision and repeatability for even the most complex dies. Equipment like 5-axis CNC mills and wire EDM machines can cut through hardened tool steels with extremely tight tolerances.

Secondary Processing: Heat treating is applied to metal parts to enhance their strength and make them more durable for their application. Grinding is used to finish parts requiring high surface quality and dimension accuracy.

Wire EDM: Wire electrical discharge machining shapes metal materials with an electrically-charged strand of brass wire. Wire EDM can cut the most intricate shapes, including small angles and contours.

Metal Stamping Design Processes

Metal stamping is a complex process that can include a number of metal forming processes blanking, punching, bending, and piercing and more. Blanking – This process is about cutting the rough outline or shape of the product. This stage is about minimizing and avoiding burrs, which can drive up the cost of your part and extend lead time. The step is where you determine hole diameter, geometry/taper, the spacing between edge-to-hole and insert the first piercing.

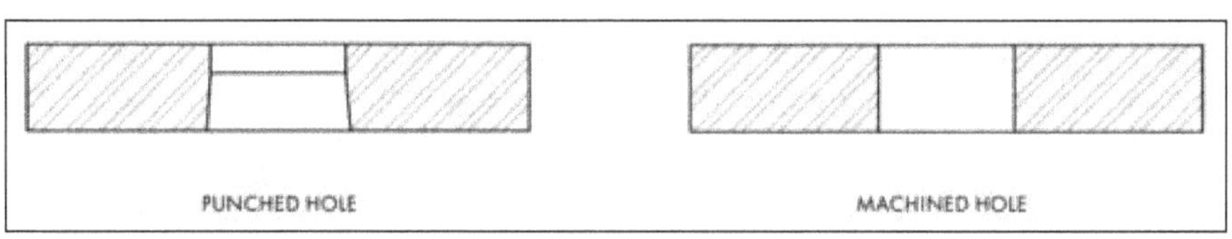

Bending: When you are designing the bends into your stamped metal part, it is important to allow for enough material — make sure to design your part and its blank so that there is enough material to perform the bend. Some important factors to remember:

- If a bend is made too close to the hole, it can become deformed.
- Notches and tabs, as well as slots, should be designed with widths that are at least 1.5x the thickness of the material. If made any smaller, they can be difficult to create due to the force exerted on punches, causing them to break.
- Every corner in your blank design should have a radius that is at least half of the material thickness.

- To minimize instances and severity of burrs, avoid sharp corners and complex cutouts when possible. When such factors cannot be avoided, be sure to note burr direction in your design so they can be taken into account during stamping.

Coining: This action is when the edges of a stamped metal part are struck to flatten or break the burr; this can create a much smoother edge in the coined area of the part geometry; this can also add additional strength to localized areas of the part and this can be utilized to avoid secondary process like deburring and grinding. Some important factors to remember:

- Plasticity and grain direction – Plasticity is the measure of permanent deformation a material undergoes when subjected to force. Metals with more plasticity are easier to form. Grain direction is important in high strength materials, such as tempered metals and stainless steel. If a bend goes along the grain of a high strength, it can be prone to cracking.

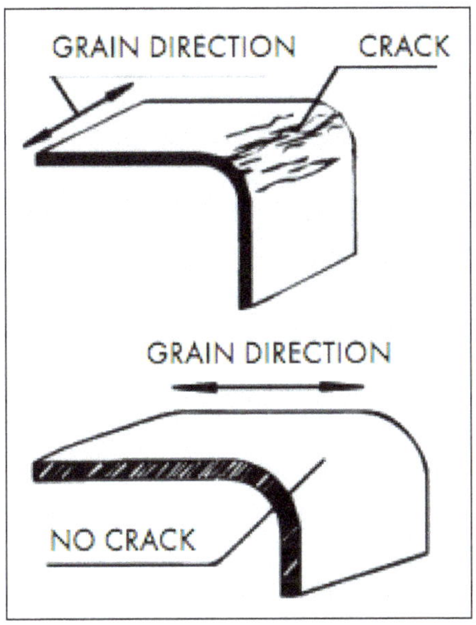

- Bend height –The overall height of a bend has minimum requirements to be formed effectively and should be at minimum, 2.5x the thickness of the material + the radius of the bend.

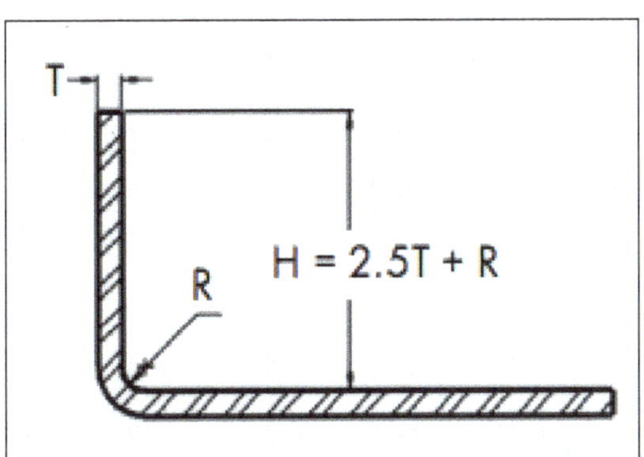

- Bend relief – Add small notches located immediately adjacent to the portion of the part to be bent — they should be a minimum of twice as wide as the material is thick, and as long as the bend radius plus material thickness.

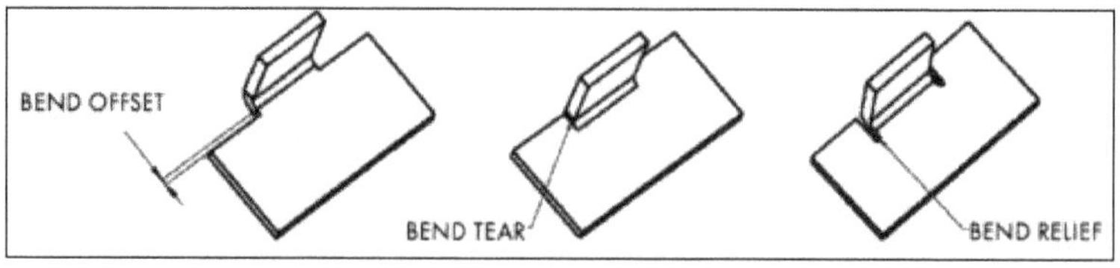

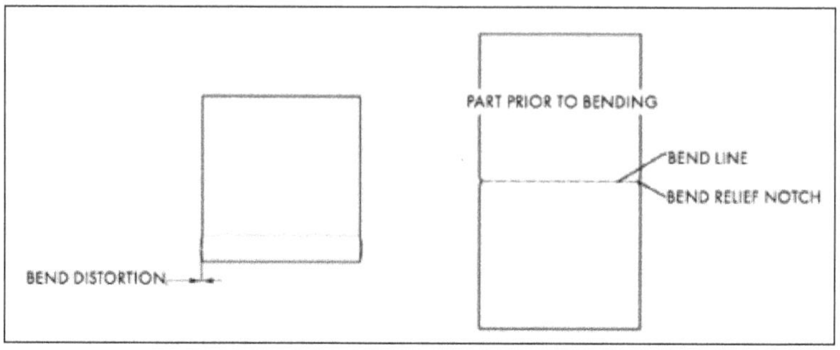

Bend Distortion/Bulge: The bulging caused by bend distortion can be as large as ½ the material thickness. As material thickness increases and bend radius decreases the distortion/bulge becomes more severe. Carrying Web and "Mismatch" Cut: This is when a very slight cut-in or bump-out on the part is required and is typically about .005" deep. This feature is not necessary when utilizing compound or transfer type tooling but is required when utilizing progressive die tooling.

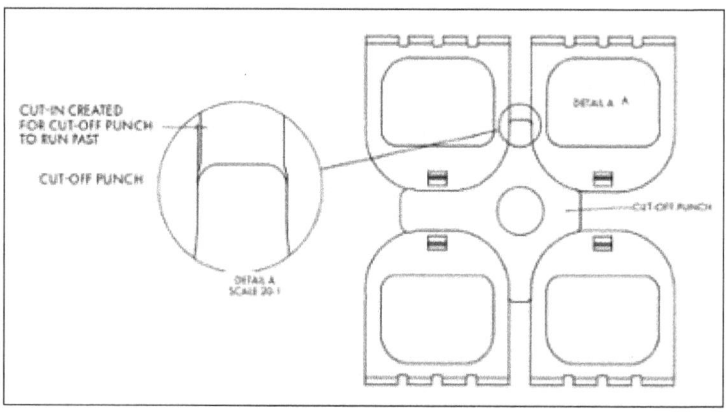

Custom Metal Stamping

Custom metal stamping describes metal forming processes that require custom tooling and techniques to produce parts specified by the customer. A wide range of industries and applications employ custom stamping processes to meet high-volume production needs and ensure all parts meet exact specifications.

HYDROFORMING

Hydroforming is a cost-effective way of shaping ductile metals such as aluminium, brass, low alloy steel, and stainless steel into lightweight, structurally stiff and strong pieces. One of the largest applications of hydroforming is the automotive industry, which makes use of the complex shapes made possible by hydroforming to produce stronger, lighter, and more rigid unibody structures for vehicles. This technique is particularly popular with the high-end sports car industry and is also frequently employed in the shaping of aluminium tubes for bicycle frames.

Hydroforming is a specialized type of die forming that uses a high pressure hydraulic fluid to press room temperature working material into a die. To hydroform aluminium into a vehicle's frame rail, a hollow tube of aluminium is placed inside a negative mold that has the shape of the desired result. High pressure hydraulic pumps then inject fluid at very high pressure inside the aluminium tube which causes it to expand until it matches the mold. The hydroformed aluminium is then removed from the mold. Hydroforming allows complex shapes with concavities to be formed, which would be difficult or impossible with standard solid die stamping. Hydroformed parts can often be made with a higher stiffness-to-weight ratio and at a lower per unit cost than traditional stamped or stamped and welded parts. Virtually all metals capable of cold forming can be hydroformed, including aluminium, brass, carbon and stainless steel, copper, and high strength alloys.

Main Process Variants

Sheet Hydroforming

This process is based on the 1950s patent for hydramolding by Fred Leuthesser, Jr. and John Fox of the Schaible Company of Cincinnati, Ohio in the United States. It was originally used in producing kitchen spouts. This was done because in addition to the strengthening of the metal, hydromolding also produced less "grainy" parts, allowing for easier metal finishing. In sheet hydroforming (SHF) there are bladder forming (where there is a bladder that contains the liquid; no liquid contacts the sheet) and hydroforming where the fluid contacts the sheet (no bladder). Bladder forming is sometimes called flexforming. Flexforming is mostly used for low volume productions, as in the aerospace field. Forming with the fluid in direct contact with the part can be done either with a male solid punch (this version is sometimes called hydro-mechanical deep drawing) or with a female solid die. In hydro-mechanical deep drawing, a work piece is placed on a draw ring (blank holder) over a male punch then a hydraulic chamber surrounds the work piece and a relatively low initial pressure seats the work piece against the punch. The punch then is raised into the hydraulic chamber and pressure is increased to as high as 100 MPa (15000 psi) which forms the part around the punch. Then the pressure is released and punch retracted, hydraulic chamber lifted, and the process is complete.

Among these techniques hydraulic bulge testing allows for an increased work hardening of sheet material by distinctive stretching operations and provides better shape accuracy for complex parts. Hence, by selecting proper material and the forming parameters for hydraulic sheet bulging study one can determine Forming Limit Curves (FLCs).

Significance

- Hydraulic bulge testing is more appropriate for sheet metal forming operations as deformation mode is bi-axial rather than uniaxial. Also it provides flow curves for the materials with extended range of plastic strain levels up to 70% before bursting occurs.
- It is helpful to generate the FLCs which will be reliable sense of reference input to the

explicit solver like LS-DYNA. These obtained FLCs are used as load curve input for such solvers for analysis.

- FLCs also serve the best for identifying the exact zone for forming operations without getting affected with localized necking and other possible defects while forming.
- Hydraulic bulge testing would be helpful to calculate the Strain hardening coefficient- "n" (i.e. Work hardening coefficient) of the material, to determine the ability of the material to be formed.
- A simple and versatile approach.
- A controlled pressure distribution over part surface during forming can be used to "control" the sheet thickness and postpone localized necking.
- The use of only single form surface tooling, which saves time and expense in the manufacture of tooling. Absence of rigid tool contact on one surface also reduces surface friction and thus surface defects, resulting in a good surface finish.

Alternative Names, Other Variants and Similar Processes

- Hydromec (Hydromechanical deep drawing).
- Aquadraw.
- Bulge forming.
- Explosive forming.
    - For large parts, explosive hydroforming can generate the forming pressure by simply exploding a charge above the part (complete with evacuated mold) which is immersed in a pool of water. The tooling can be much cheaper than what would be required for any press-type process. The hydroforming-into-a-mold process also works using only a shock wave in air as the pressuring medium. Particularly when the explosives are close to the workpiece, inertia effects make the result more complicated than forming by hydrostatic pressure alone.
- Rubber pad forming.

Tube Hydroforming

In tube hydroforming (THF) there are two major practices: high pressure and low pressure. With the high pressure process the tube is fully enclosed in a die prior to pressurization of the tube. In low pressure the tube is slightly pressurized to a fixed volume during the closing of the die (this used to be called the Variform process). Historically, the process was patented in the '50s, but it was industrially spread in the '70s for the production of large T-shaped joints for the oil & gas industry.Today it is mostly used in the automotive sector, where many industrial applications can be found. It is also a method of choice for several tubular members of bicycles. In tube hydroforming pressure is applied to the inside of a tube that is held by dies with the desired cross sections and forms. When the dies are closed, the tube ends are sealed by axial punches and the tube is filled with hydraulic fluid. The internal pressure can go up to a few thousand bars and it causes the tube to calibrate against the dies. The fluid is injected into the tube through one of the two axial punches. Axial punches are movable and their action is required to provide axial compression and to feed material towards the center of the bulging tube. Transverse counterpunches may also be incorporated in the forming die in order to form protrusions with small diameter/length ratio. Transverse counter punches may also be used to punch holes in the work piece at the end of the forming process.

Designing the process has in the past been a challenging task, since initial analytical modeling is possible only for limited cases. Advances in FEA and FEM in recent years has enabled hydroform processes to be more widely engineered for varieties of parts and materials. Often FEM simulations must be performed in order to find a feasible process solution and to define the correct loading curves: pressure vs. time and axial feed vs. time. In the case of more complex tube hydroformed parts the tube must be pre-bent prior to loading into the hydroforming die. Bending is done sequentially along the length of the tube, with the tube being bent around bending discs (or dies) as the tube length is fed in. Bending can be done with or without mandrels. This additional complexity of process further increases the reliance on FEM for designing and evaluating manufacturing processes. The feasibility of a hydroforming process must take into consideration the initial tube material properties and its potential for variation, along with the bending process, hydraulic pressure throughout the forming process, in inclusion of axial feed or not, in order to predict metal formability.

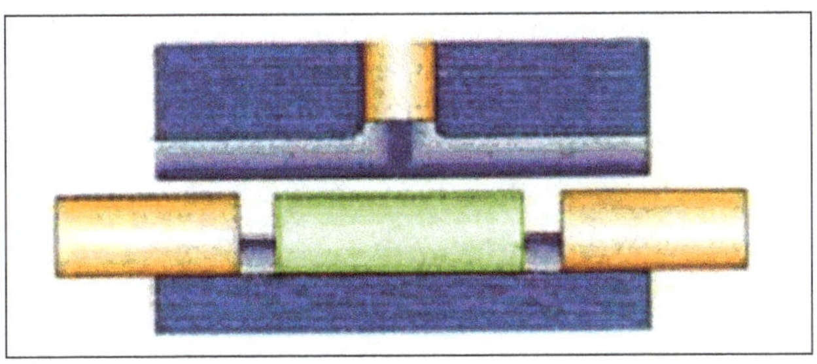

Process sequence in tube hydroforming of a t-shape with counterpunch.

Typical Tools

Tools and punches can be interchanged for different part requirements. One advantage of hydroforming is the savings on tools. For sheet metal only a draw ring and punch (metalworking) or male die is required. Depending on the part being formed, the punch can be made from epoxy, rather than metal. The bladder of the hydroform itself acts as the female die eliminating the need to fabricate it. This allows for changes in material thickness to be made with usually no necessary changes to the tool. However, dies must be highly polished and in tube hydroforming a two-piece die is required to allow opening and closing.

Geometry Produced

Another advantage of hydroforming is that complex shapes can be made in one step. In sheet hydroforming (SHF) with the bladder acting as the male die almost limitless geometries can be produced. However, the process is limited by the very high closing force required in order to seal the dies, especially for large panels and thick hard materials. Small concave corner radii are difficult to be completely calibrated, i.e. filled, because too large a pressure would be required. in fact, the die closing force can be very high, both in tube and sheet hydroforming and may easily overcome the maximum tonnage of the forming press. In order to keep the die closing force under prescribed limits, the maximum internal fluid pressure must be limited. This reduces the calibration abilities of the process, i.e. it reduces the possibility of forming parts with small concave radii. Limits of the SHF process are due to risks of excessive thinning, fracture, wrinkling and are strictly related to the material formability and to a proper selection of process parameters (e.g. hydraulic pressure vs. time curve). Tube hydroforming (THF) can produce many geometric options as well, reducing the need for tube welding operations. Similar limitations and risks can be listed as in SHF; however, the maximum closing force is seldom a limiting factor in THF.

Tolerances and Surface Finish

Hydroforming is capable of producing parts within tight tolerances including aircraft tolerances where a common tolerance for sheet metal parts is within 0.76 mm (1/30th of an inch). Metal hydroforming also allows for a smoother finish as draw marks produced by the traditional method of pressing a male and female die together are eliminated.

While springback has long been a topic of discussion for sheet metal forming operations it has been far less of a topic of research for tube hydroforming. This may in part be a result of the relatively low levels of springback naturally occurring when deforming the tubes into their closed section geometries. Tube Hydroformed sections by the nature of their closed sections are very rigid and do not display high degrees of elastic deformation under load. For this reason it is likely that negative residual stress induced during tube hydroforming might be insufficient to deform the part elastically after the completion of forming. However, as more and more tubular parts are being manufactured using high strength steel and advanced high strength steel parts, springback must be accounted for in the design and manufacture of closed section tube hydroformed parts.

Examples:

Notable examples include:

SHF

- Satellite antennas up to 6 meters in diameter, such as those used in the Allen Telescope Array.
- Lighting fixture housing and reflector.

THF

- The brass tube of Yamaha saxophones.
- The process has become popular for the manufacture of aluminium bicycle frames. The earliest commercially manufactured one being that of the Giant Manufacturing Revive bicycle first marketed in 2003.

- Many motor vehicles have major components manufactured using this technology, for example:
    - The technique is widely used in the manufacture of engine cradles. The first mass-produced one was for the Ford Contour and Mystique in 1994. Others from a long list include the Pontiac Aztek, the Honda Accord and the perimeter frame around the Harley Davidson V-Rod motorcycle's engine.
    - As well as engine cradles, the main automotive applications for hydroforming are suspension, radiator supports and instrument-panel support beams. The 1994 Buick Regal and Oldsmobile Cutlass had hydro-formed instrument panel beams. The first mass-produced automotive component was in 1990 with the instrument panel support beam for the Chrysler minivan.
    - Various vehicle bodies and body components, the earliest mass-produced one being the 1997 Chevrolet Corvette. A selection from many examples are the current versions of the three major United States pickup trucks—the Ford F-150, Chevrolet Silverado, and Ram—which all have hydroformed frame rails, 2006 Pontiac Solstice and the steel frame inside the John Deere HPX Gator Utility Vehicle.

DECAMBERING

Decambering is the metalworking process of removing camber, or horizontal bend, from strip shaped materials. The material may be finite length sections or continuous coils. Decambering resembles flattening or levelling processes, but deforms the material edge (left or right) instead of the face (up or down) of the strip.

Processes

Decambering may be accomplished by a choice of several processes:

The material may be "stretchformed" or stretched typically 1-2% elongation past its elastic limit, such that the camber is strained out, and the strip relaxes straight. Typically this is accomplished by routing the strip through two sets of 180 degree or more wrap driven "bridle rolls" with the second rolls rotating faster than the first. Cut strip can be gripped each end then stretched on a static rig with one end being mechanically pulled. The "stretchforming" process is most suited to thin strip.

Wider material can be flat rolled with the rolls tilting such that the "short" edge is lengthened, with thickness slightly reduced, to equal the "long" edge. The material is then straight if properly processed. Several different types of these machines are available, ranging from position controlled to pressure controlled machines. Also, machines are available with automatic self checking features. If the material has residual stress as in toothed bandsaw blades, the rolls must not uniformly apply cross strain, as the inherent strain will distort the strip. In these cases, rolls exert corrective strain in a linear narrow strip inboard of the edge each side. In the case of sawblades with teeth on, the corrective squeezed strip on the toothed edge is located along the tooth gullet, and corrective squeeze on the toothed side must be about 2.5 times the corrective force on the plain edge.

If the strip is narrow and thick enough, side rolls with guide grooves can be used to bend the material laterally to remove the camber. There are many patented designs of these machines, with the best engaging the material on both edges of the strip over the worked portion, preventing the material from twisting or "cheating" the corrective lateral bending.

Note that camber is often caused by slitting, where slitting knives extend the edge being slit. Outer edges left untreated usually have a shorter length than the inner edges being formed by slitting. The outer mults from slitting will therefore usually have camber, while inner mults frequently will not. Decamberers are cost effective when material is not laterally straight enough to suit a given process, and needs upgrading. Cambered material is usually lower in cost than straight material, and decamberers offer savings in these cases. Various types of decambering machines cater to wide or narrow materials.

EXPLOSIVE FORMING

Explosive forming is a metalworking technique in which an explosive charge is used instead of a punch or press. It can be used on materials for which a press setup would be prohibitively large or require an unreasonably high pressure, and is generally much cheaper than building a large enough and sufficiently high-pressure press; on the other hand, it is unavoidably an individual job production process, producing one product at a time and with a long setup time.

Various Approaches

There are various approaches; one is to place metal plate over a die, with the intervening space evacuated by a vacuum pump, place the whole assembly underwater, and detonate a charge at an appropriate distance from the plate. For complicated shapes, a segmented die can be used to produce in a single operation a shape that would require many manufacturing steps, or to be manufactured in parts and welded together with an accompanying loss of strength at the welds. There is often some degree of work hardening from the explosive-forming process, particularly in mild steel.

Tooling

Tooling can be made out of fiberglass for short-run applications, out of concrete for large parts at medium pressures, or out of ductile iron for high-pressure work; ideally the tooling should have higher yield strength than the material that is being formed, which is a problem since the technique is usually only considered for material which is itself very hard to work.

Explosive Forming of Vacuum Tube Anode (Plate) Materials

In the late 1950s, the General Electric company developed an application for five-layer sheet metal composites that had been created using the explosive forming process. GE engineers used this innovative composite material to produce multi-layer vacuum tube anodes (aka "plates") with superior heat transfer characteristics. This characteristic allowed GE to build significantly higher power vacuum tubes from existing designs without expensive engineering, design, and tooling changes, providing a substantial competitive market advantage to GE in the burgeoning Hi-Fi amplifier market.

In January 1960 it was reported in contemporary GE technical literature that this five-layer material was the design breakthrough which made possible the new 6L6GC. The 6L6GC was a 6L6 variant able to dissipate 26% more power compared to the otherwise identically constructed 6L6GB. these increases were made possible by the application of the improved multi-layer plate material.

GE sourced this material from a Texas-based firm (Texas Instruments) which is reported to be the source of the explosively forged five-layer raw material specified by General Electric engineers. This manufacturer used explosive sheet metal forging processes previously developed for another customer (possibly the U.S. Navy?) The explosively formed dissimilar materials had substantially improved evenness of heat transfer thanks to the copper center layer.

GE engineers quickly saw the potential for improved heat transfer characteristics in several already popular pentode and beam tetrode vacuum tube designs, including the 6L6GB, the 7189, and eventually the 6550. The application of the five-layer (Al-Fe-Cu-Fe-Al) material to anode manufacture solved the problem of irregular heat buildup at high power levels in the anode plates of power pentodes, tetrodes, and triodes. This irregular heat buildup leads to physical distortion of the tube's plate. if allowed to continue, this spot overheating eventually results in warpage which allows physical contact and subsequent short circuits between the plate, grids, and beam formers in the tube. Such contact shorts destroy the tube.

General Electric's novel application of this innovative composite led to the creation of the 7189A variant, released in late 1959, along with the 6L6GC and other variants. By 1969, the 6550A variant had also been developed to take advantage of explosively forged composites. GE's application allowed for improved power levels in a number of already popular tube designs, an innovation which helped pave the way for substantially higher power vacuum tube stereo and musical instrument amplifiers in the 1960s and early 1970s.

DRAWING

Drawing is a metalworking process which uses tensile forces to stretch metal or glass. As the metal is drawn (pulled), it stretches thinner, into a desired shape and thickness. Drawing is classified in two types: sheet metal drawing and wire, bar, and tube drawing. The specific definition for sheet metal drawing is that it involves plastic deformation over a curved axis. For wire, bar, and tube drawing the starting stock is drawn through a die to reduce its diameter and increase its length. Drawing is usually done at room temperature, thus classified a cold working process, however it may be performed at elevated temperatures to hot work large wires, rods or hollow sections in order to reduce forces.

Drawing differs from rolling in that the pressure of drawing is not transmitted through the turning action of the mill but instead depends on force applied locally near the area of compression. This means the amount of possible drawing force is limited by the tensile strength of the material, a fact that is particularly evident when drawing thin wires.

The starting point of cold drawing is hot rolled stock of a suitable size.

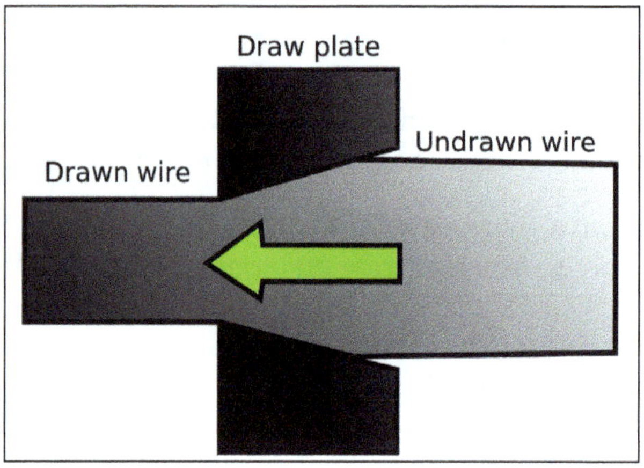

The basic drawing process for a wire, bar or tube.

Processes

Metal: Sheet Metal

The success of forming is in relation to two things, the flow and stretch of material. As a die forms a shape from a flat sheet of metal, there is a need for the material to move into the shape of the die. The flow of material is controlled through pressure applied to the blank and lubrication applied to the die or the blank. If the form moves too easily, wrinkles will occur in the part. To correct this, more pressure or less lubrication is applied to the blank to limit the flow of material and cause the material to stretch or set thin. If too much pressure is applied, the part will become too thin and break. Drawing metal requires finding the correct balance between wrinkles and breaking to achieve a successful part.

Deep Drawing and Shallow Drawing

Sheet metal drawing becomes *deep drawing* when the workpiece is drawing longer than its diameter. It is common that the workpiece is also processed using other forming processes, such as piercing, ironing, necking, rolling, and beading. In *shallow drawing*, the depth of drawing is less than the smallest dimension of the hole.

Bar, Tube and Wire

Bar, tube, and wire drawing all work upon the same principle: the starting stock drawn through a die to reduce the diameter and increase the length. Usually the die is mounted on a draw bench. The end of the workpiece is reduced or pointed to get the end through the die. The end is then placed in grips and the rest of the workpiece is pulled through the die. Steels, copper alloys, and aluminium alloys are common materials that are drawn.

Drawing can also be used to cold-form a shaped cross-section. Cold drawn cross-sections are more precise and have a better surface finish than hot extruded parts. Inexpensive materials can be used instead of expensive alloys for strength requirements, due to work hardening.

Bar Drawing

Bars or rods that are drawn cannot be coiled therefore straight-pull draw benches are used. Chain drives are used to draw workpieces up to 30 m (98 ft). Hydraulic cylinders are used for shorter length workpieces.

The reduction in area is usually restricted to between 20 and 50%, because greater reductions would exceed the tensile strength of the material, depending on its ductility. To achieve a certain size or shape multiple passes through progressively smaller dies or intermediate anneals may be required.

Tube Drawing

Tube drawing is very similar to bar drawing, except the beginning stock is a tube. It is used to decrease the diameter, improve surface finish and improve dimensional accuracy. A mandrel may or may not be used depending on the specific process used.

Wire Drawing

This technique has long been used to produce flexible metal wire by drawing the material through a series of dies of decreasing size. These dies are manufactured from a number of materials, the most common being tungsten carbide and diamond.

The Cold Drawing Process for Steel Bars and Wire

- Coating: The surface of the bar or coil is coated with a drawing lubricant to aid cold drawing.
- Pointing: Several inches of the lead ends of the bar or coil are reduced in size by swaging or extruding so that it can pass freely through the drawing die. This is done because the die opening is always smaller than the original bar or coil section size.
- Cold Drawing, ProcessDrawing: In this process, the material being drawn is at room temperature (i.e. Cold-Drawn). The pointed/reduced end of the bar or coil, which is smaller than the die opening, is passed through the die where it enters a gripping device of the drawing machine. The drawing machine pulls or draws the remaining unreduced section of the bar or coil through the die. The die reduces the cross section of the original bar or coil, shapes the profile of the product and increases the length of the original product.
- Finished Product: The drawn product, which is referred to as Cold Drawn or Cold Finished, exhibits a bright and/or polished finish, increased mechanical properties, improved machining characteristics and precise and uniform dimensional tolerances.
- Multi-Pass Drawing: The cold drawing of complex shapes/profiles may require that each bar/coil be drawn several times in order to produce the desired shape and tolerances. This

process is called multi-pass drawing and involves drawing through smaller and smaller die openings. Material is generally annealed between each drawing pass to remove cold work and to increase ductility.

- Annealing: This is a thermal treatment generally used to soften the material being drawn, to modify the microstructure, the mechanical properties and the machining characteristics of the steel and/or to remove internal stresses in the product. Depending on the desired characteristics of the finished product, annealing may be used before, during (between passes) or after the cold drawing operation, depending on material requirements.

Plastics

Plastic drawing, sometimes referred to as *cold drawing*, is the same process as used on metal bars, but applied to plastics.

Plastic drawing is primarily used in manufacturing plastic fibers. The process was discovered by Julian W. Hill in 1930 while trying to make fibers from an early polyester. It is performed after the material has been "spun" into filaments; by extruding the polymer melt through pores of a spinneret. During this process, the individual polymer chains tend to somewhat align because of viscous flow. These filaments still have an amorphous structure, so they are drawn to align the fibers further, thus increasing crystallinity, tensile strength and stiffness. This is done on a draw twister machine.

For nylon, the fiber is stretched four times its spun length. The crystals formed during drawing are held together by hydrogen bonds between the amide hydrogens of one chain and the carbonyl oxygens of another chain.

Glass

A similar process is applicable to glass.

RUBBER PAD FORMING PROCESS

The rubber pad forming process is also called as "guerin" process. The die is placed on lower bed and top of that soft die of rubber is placed. The rubber die is forced on the blank and deform the blank into require shape. During the rubber pad forming process, the punch deforms the blank and apply large pressure (100 Mpa) over the blank. The punch is act universal tool, which is used to provide various different parts. The rubber pad forming sheet metal parts having without stress.

Solid Rubber Pad Forming

It is easy and requires application of rigid container fillet with solid rubber pad.

The blank of sheet metal work piece is placed on the die. Before the forming process, the blanks are precut to their final dimension.

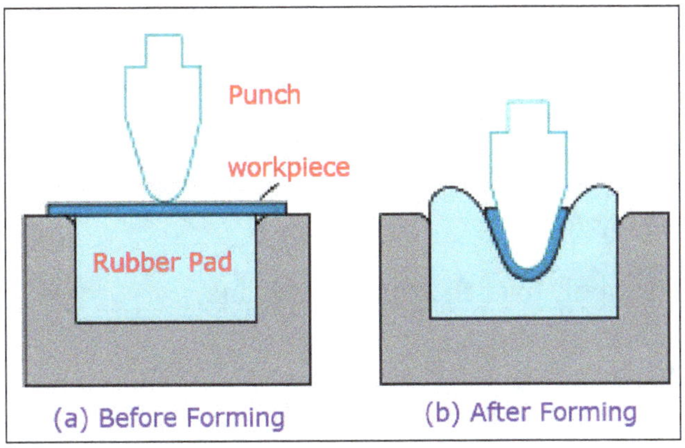

Solid rubber pad forming process.

The punch become downward the presses the work blank again the rubber pad. So, the work metal is forced on the punch. So the form of punch is obtained on the sheet metal. The rubber pad in one side of blank to minimize the friction between the die and blank, so the damage and wrinkling of blank will be avoided during the pad forming process. The male or female dies are used. The male die is used to make a flanged parts. The female die is used for making a deep drawn parts, that complexity will be less. After complete forming the part have to be trimming to their require final dimension.

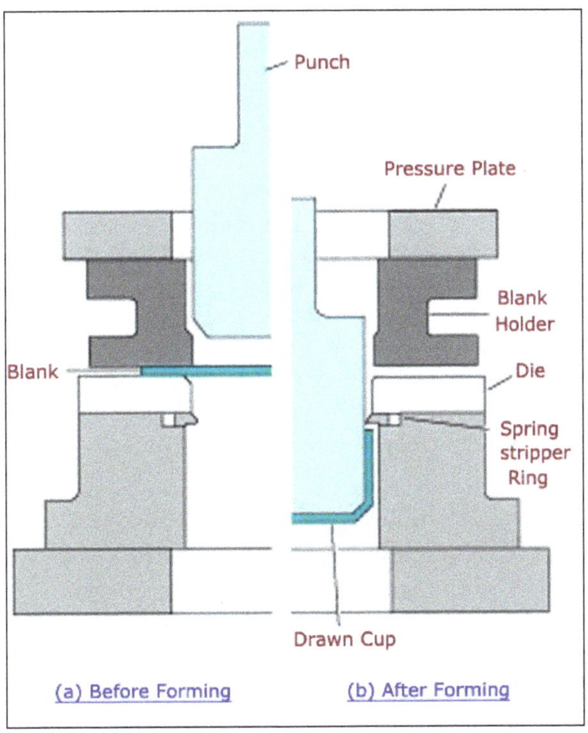

Rubber pad forming sheet metal of Flanged parts.

Advantages of Rubber Pad Forming Process

- The soft rubber placed on the top punch and will not damage or scratch the sheet metal.
- Accurate and time consuming of die is not necessary.

- That require only one punch to manufacture of sheet metal.
- The special surface layer such as clad layers.
- Any of shape can be formed in one process cycle.
- Tooling cost will be low.
- Uniform pressure to apply over the surface of blank.

Disadvantages of Rubber Pad Forming Sheet Metal

- Require long time.
- Limit the number of parts.
- Large force required.
- Soft tool is deformed during the process.
- Complex shape part formed limited.
- The soft tool is wear also.

Application of Rubber Pad Forming Sheet Metal

- Flanged parts and deep drawn parts.
- Steel alloys.
- Air craft industry.
- Aluminium alloy.

HEMMING

Hemming is a forming operation in which the edges of the sheet are folded or folded over another part in order to achieve a tight fit. Normally hemming operations are used to connect parts together, to improve the appearance of a part and to reinforce part edges.

In car part production, hemming is used in assembly as a secondary operation after deep drawing, trimming and flanging operations to join two sheet metal parts (outer and inner) together. Typical parts for this type of assembly are hoods, doors, trunk lids and fenders.

The accuracy of the hemming operation is very important since it affects the appearance of the surface and surface quality. Material deformations, which occur during the hemming process, can lead to dimensional variations and other defects in parts. Typical hemming defects are splits or wrinkles in the flange, material overlaps in the corner areas or material roll-in. This is why it is important to use simulation tools in order to, on the one hand, better understand the hemming process and, on the other hand, significantly reduce the number of "trial and error" loops during tryout und production.

There are various types of hemming operations:

- Conventional die hemming.
- Roll hemming.

In conventional die hemming, the flange is folded over the entire length with a hemming tool.

In roll hemming, the hemming roller is guided by an industrial robot to form the flange.

Conventional Die Hemming

Conventional die hemming is suitable for mass production. In die hemming, the flange is folded over the entire length with a hemming tool. Normally the actual hemming is a result of a forming operation in which the flange is formed with a hemming tool after the drawing and trimming operations have been completed. The formed flange is then hemmed in several process steps. These steps include, for example, the pre-hemming and final hemming depending on the respective opening angle of the flange. Production plants for conventional die hemming are very expensive, but the cycle times are very low.

Roll Hemming

Roll hemming is carried out incrementally with a hemming roller. An industrial robot guides the hemming roller and forms the flange. Roll hemming operation can also be divided into several pre-hemming and final hemming process steps. Roll hemming is very flexible to use and tool costs are significantly lower as compared to those of conventional die hemming. However, the cycle times are much higher since the hemming is realized using a hemming roller which follows a defined path.

References

- Spinning the Elements – Cold Drawing, Chemical Heritage Foundation, archived from the original on 2001-05-04, retrieved 2008-11-13
- Hemming, glossary: autoform.com, Retrieved 19 June, 2019
- Strano, Matteo; Jirathearanat, Suwat; Shr, Shiuan-Guang; Altan, Taylan (2004). "Virtual process development in tube hydroforming". Journal of Materials Processing Technology. 146 (1): 130–136. Doi:10.1016/S0924-0136(03)00853-7
- Rubber-pad-forming-process: techminy.com, Retrieved 20 July, 2019
- Asnafi, Nader (1999). "Analytical modelling of tube hydroforming". Thin-Walled Structures. 34 (4): 295–330. Doi:10.1016/S0263-8231(99)00018-X

Machining

6

CHAPTER

Machining refers to the process in which a piece of raw material is cut into a desired final shape and size by a controlled material-removal process. Some of the machining processes are milling, turning, drilling, broaching, mass finishing, honing and hobbing. These diverse machining processes have been thoroughly discussed in this chapter.

Machining is a manufacturing term encompassing a broad range of technologies and techniques. It can be roughly defined as the process of removing material from a workpiece using power-driven machine tools to shape it into an intended design. Most metal components and parts require some form of machining during the manufacturing process. Other materials, such as plastics, rubbers, and paper goods, are also commonly fabricated through machining processes.

Types of Machining Tools

There are many types of machining tools, and they may be used alone or in conjunction with other tools at various steps of the manufacturing process to achieve the intended part geometry. The major categories of machining tools are:

- Boring tools: These are typically used as finishing equipment to enlarge holes previously cut into the material.
- Cutting tools: Devices such as saws and shears are typical examples of cutting implements. They are often used to cut material with predetermined dimensions, such as sheet metal, into a desired shape.
- Drilling tools: This category consists of two-edged rotating devices that create round holes parallel to the axis of rotation.
- Grinding tools: These instruments apply a rotating wheel to achieve a fine finish or to make light cuts on a workpiece.
- Milling tools: A milling tool employs a rotating cutting surface with several blades to create non-circular holes or cut unique designs out of the material.
- Turning tools: These tools rotate a workpiece on its axis while a cutting tool shapes it to form. Lathes are the most common type of turning equipment.

Types of Burning Machining Technologies

Welding and burning machine tools use heat to shape a workpiece. The most common types of welding and burning machining technologies include:

- Laser cutting: A laser machine emits a narrow, high-energy beam of light that effectively melts, vaporizes, or burns material. CO2 and Nd:YAG lasers are the most common types used in machining. The laser cutting process is well-suited for shaping steel or etching patterns into a piece of material. Its benefits include high-quality surface finishes and extreme cutting precision.
- Oxy-fuel cutting: Also known as gas cutting, this machining method employs a mixture of fuel gases and oxygen to melt and cut away material. Acetylene, gasoline, hydrogen, and propane frequently serve as gas media due to their high flammability. This method's benefits include high portability, a low dependence on primary power sources, and the ability to cut thick or hard materials, such as sturdy steel grades.
- Plasma cutting: Plasma torches fire an electrical arc to transform inert gas into plasma. This plasma reaches extremely elevated temperatures and is applied to the workpiece at high speed to melt away unwanted material. The process is often used on electrically conductive metals that require a precise cut width and minimal prep time.

Types of Erosion Machining Technologies

While burning tools apply heat to melt excess stock, erosion machining devices use water or electricity to erode material off the workpiece. The two main types of erosion machining technologies are:

- Water jet cutting: This process uses a high-pressurized stream of water to cut through material. Abrasive powder may be added to the water stream to facilitate erosion. Water jet cutting is typically used on materials that can suffer damage or deformation from a heat affected zone.
- Electric discharge machining (EDM): Also known as spark machining, this process uses electric arcing discharges to create micro-craters that rapidly result in complete cuts. EDM is used in applications requiring complex geometrical shapes in hard materials and at close tolerances. EDM requires the base material to be electrically conductive, which limits its use to ferrous alloys.

CNC Machining

Computer numerical control, or CNC, machining is a computer-aided technique that can be used in conjunction with a broad range of equipment. It requires software and programming, usually in the G-code language, to guide a machining tool in shaping the workpiece according to preset parameters. As opposed to manually guided methods, CNC machining is an automated process. Some of its benefits include:

- High production cycles: Once the CNC machine has been properly coded, it usually needs minimal maintenance or downtime, allowing for a faster production rate.
- Low manufacturing costs: Due to its turnover speed and low manual labor requirements, CNC machining can be a cost-efficient process, particularly for high-volume production runs.

- Uniform production: CNC machining is typically precise and yields a high level of design consistency among its products.

Precision Machining

Any machining process that requires unusually small cutting tolerances (between 0.013 mm and 0.0005 mm, as a rule of thumb) or surface finishes finer than 32T may be considered a form of precision machining. Like CNC machining, precision machining can be applied to a wide number of fabrication methods and tools. Factors such as stiffness, damping, and geometric accuracy can influence the exactness of a precision tool's cut. Motion control and the machine's ability to respond at rapid feed rates are also important in precision machining applications.

MILLING

Milling is the process of machining using rotary cutters to remove material by advancing a cutter into a workpiece. This may be done varying direction on one or several axes, cutter head speed, and pressure. Milling covers a wide variety of different operations and machines, on scales from small individual parts to large, heavy-duty gang milling operations. It is one of the most commonly used processes for machining custom parts to precise tolerances.

Milling can be done with a wide range of machine tools. The original class of machine tools for milling was the milling machine (often called a mill). After the advent of computer numerical control (CNC) in the 1960s, milling machines evolved into *machining centers:* milling machines augmented by automatic tool changers, tool magazines or carousels, CNC capability, coolant systems, and enclosures. Milling centers are generally classified as vertical machining centers (VMCs) or horizontal machining centers (HMCs).

The integration of milling into turning environments, and vice versa, began with live tooling for lathes and the occasional use of mills for turning operations. This led to a new class of machine tools, multitasking machines (MTMs), which are purpose-built to facilitate milling and turning within the same work envelope.

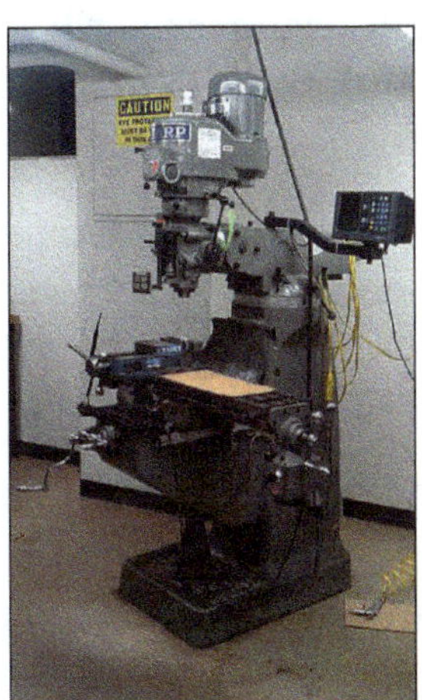

Full view of a 3-axis clone of a Bridgeport-style vertical milling machine.

Process

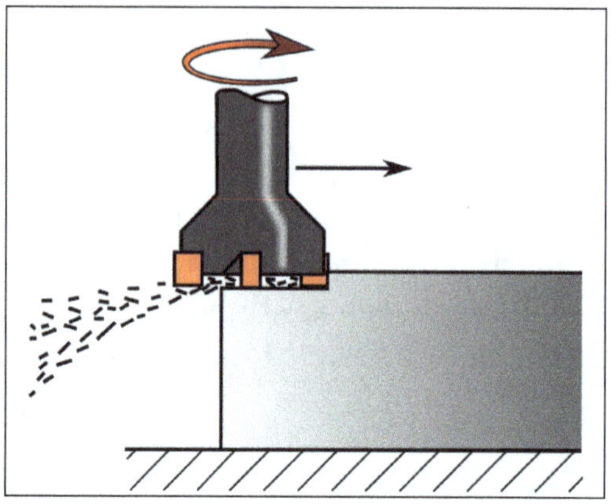

Face milling process (cutter rotation axis is vertical - 0° inclination relative to tool axis).

Milling is a cutting process that uses a milling cutter to remove material from the surface of a workpiece. The milling cutter is a rotary cutting tool, often with multiple cutting points. As opposed to drilling, where the tool is advanced along its rotation axis, the cutter in milling is usually moved perpendicular to its axis so that cutting occurs on the circumference of the cutter. As the milling cutter enters the workpiece, the cutting edges (flutes or teeth) of the tool repeatedly cut into and exit from the material, shaving off chips (swarf) from the workpiece with each pass. The cutting action is shear deformation; material is pushed off the workpiece in tiny clumps that hang together to a greater or lesser extent (depending on the material) to form chips. This makes metal cutting somewhat different (in its mechanics) from slicing softer materials with a blade.

The milling process removes material by performing many separate, small cuts. This is accomplished by using a cutter with many teeth, spinning the cutter at high speed, or advancing the material through the cutter slowly; most often it is some combination of these three approaches. The speeds and feeds used are varied to suit a combination of variables. The speed at which the piece advances through the cutter is called feed rate, or just feed; it is most often measured in length of material per full revolution of the cutter.

There are two major classes of milling process:

- In face milling, the cutting action occurs primarily at the end corners of the milling cutter. Face milling is used to cut flat surfaces (faces) into the workpiece, or to cut flat-bottomed cavities.
- In peripheral milling, the cutting action occurs primarily along the circumference of the cutter, so that the cross section of the milled surface ends up receiving the shape of the cutter. In this case the blades of the cutter can be seen as scooping out material from the work piece. Peripheral milling is well suited to the cutting of deep slots, threads, and gear teeth.

Milling Cutters

Many different types of cutting tools are used in the milling process. Milling cutters such as endmills may have cutting surfaces across their entire end surface, so that they can be drilled into the workpiece (plunging). Milling cutters may also have extended cutting surfaces on their sides to allow for peripheral milling. Tools optimized for face milling tend to have only small cutters at their end corners.

The cutting surfaces of a milling cutter are generally made of a hard and temperature-resistant material, so that they wear slowly. A low cost cutter may have surfaces made of high speed steel. More expensive but slower-wearing materials include cemented carbide. Thin film coatings may be applied to decrease friction or further increase hardness.

There are cutting tools typically used in milling machines or machining centres to perform milling operations (and occasionally in other machine tools). They remove material by their movement within the machine (e.g., a ball nose mill) or directly from the cutter's shape (e.g., a form tool such as a hobbing cutter).

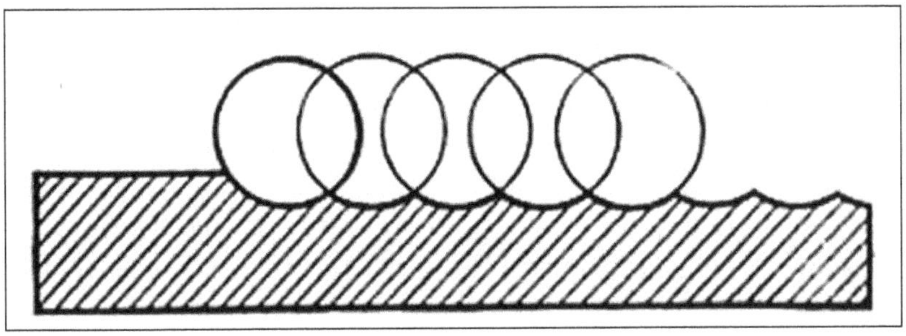

A diagram of revolution ridges on a surface milled by the side of the cutter, showing the position of the cutter for each cutting pass and how it corresponds with the ridges (cutter rotation axis is perpendicular to image plane).

As material passes through the cutting area of a milling machine, the blades of the cutter take swarfs of material at regular intervals. Surfaces cut by the side of the cutter (as in peripheral milling) therefore always contain regular ridges. The distance between ridges and the height of the ridges depend on the feed rate, number of cutting surfaces, the cutter diameter. With a narrow cutter and rapid feed rate, these revolution ridges can be significant variations in the surface finish.

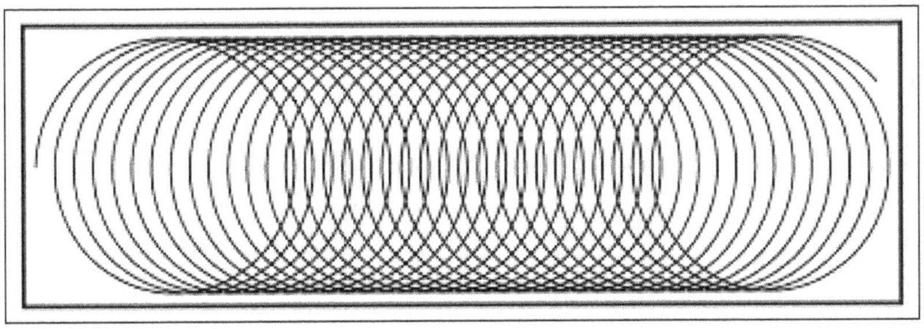

Trochoidal marks, characteristic of face milling.

The face milling process can in principle produce very flat surfaces. However, in practice the result always shows visible trochoidal marks following the motion of points on the cutter's end face. These revolution marks give the characteristic finish of a face milled surface. Revolution marks can have significant roughness depending on factors such as flatness of the cutter's end face and the degree of perpendicularity between the cutter's rotation axis and feed direction. Often a final pass with a slow feed rate is used to improve the surface finish after the bulk of the material has been removed. In a precise face milling operation, the revolution marks will only be microscopic scratches due to imperfections in the cutting edge.

Heavy gang milling of milling machine tables.

Gang milling refers to the use of two or more milling cutters mounted on the same arbor (that is, ganged) in a horizontal-milling setup. All of the cutters may perform the same type of operation, or each cutter may perform a different type of operation. For example, if several workpieces need a slot, a flat surface, and an angular groove, a good method to cut these (within a non-CNC context) would be gang milling. All the completed workpieces would be the same, and milling time per piece would be minimized.

Gang milling was especially important before the CNC era, because for duplicate part production, it was a substantial efficiency improvement over manual-milling one feature at an operation, then changing machines (or changing setup of the same machine) to cut the next op. Today, CNC mills with automatic tool change and 4- or 5-axis control obviate gang-milling practice to a large extent.

Equipment

Milling is performed with a milling cutter in various forms, held in a collett or similar which, in turn, is held in the spindle of a milling machine.

Types and Nomenclature

Mill orientation is the primary classification for milling machines. The two basic configurations are vertical and horizontal. However, there are alternative classifications according to method of control, size, purpose and power source.

Mill Orientation

Vertical Mill

In the vertical mill the spindle axis is vertically oriented. Milling cutters are held in the spindle and rotate on its axis. The spindle can generally be extended (or the table can be raised/lowered, giving the same effect), allowing plunge cuts and drilling. There are two subcategories of vertical mills: the bed mill and the turret mill.

- A turret mill has a stationary spindle and the table is moved both perpendicular and parallel to the spindle axis to accomplish cutting. The most common example of this type is the

Bridgeport, described below. Turret mills often have a quill which allows the milling cutter to be raised and lowered in a manner similar to a drill press. This type of machine provides two methods of cutting in the vertical (Z) direction: by raising or lowering the quill, and by moving the knee.

- In the bed mill, however, the table moves only perpendicular to the spindle's axis, while the spindle itself moves parallel to its own axis.

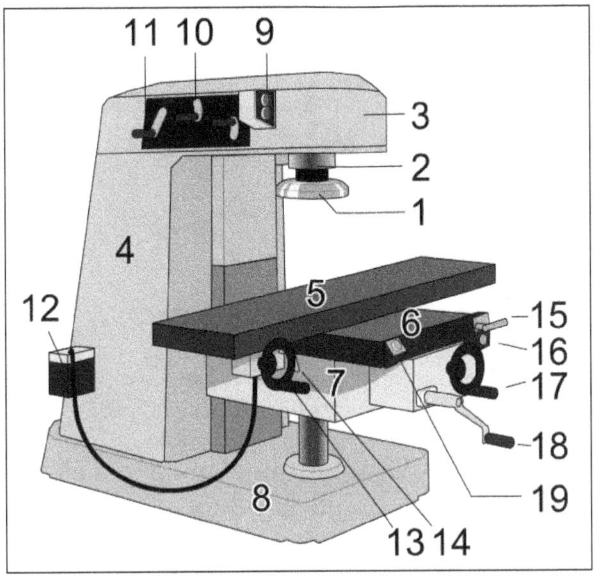

Vertical milling machine. 1: milling cutter 2: spindle 3: top slide or overarm 4: column 5: table 6: Y-axis slide 7: knee 8: base.

Turret mills are generally considered by some to be more versatile of the two designs. However, turret mills are only practical as long as the machine remains relatively small. As machine size increases, moving the knee up and down requires considerable effort and it also becomes difficult to reach the quill feed handle (if equipped). Therefore, larger milling machines are usually of the bed type.

A third type also exists, a lighter machine, called a mill-drill, which is a close relative of the vertical mill and quite popular with hobbyists. A mill-drill is similar in basic configuration to a small drill press, but equipped with an X-Y table. They also typically use more powerful motors than a comparably sized drill press, with potentiometer-controlled speed and generally have more heavy-duty spindle bearings than a drill press to deal with the lateral loading on the spindle that is created by a milling operation. A mill drill also typically raises and lowers the entire head, including motor, often on a dovetailed vertical, where a drill press motor remains stationary, while the arbor raises and lowers within a driving collar. Other differences that separate a mill-drill from a drill press may be a fine tuning adjustment for the Z-axis, a more precise depth stop, the capability to lock the X, Y or Z axis, and often a system of tilting the head or the entire vertical column and powerhead assembly to allow angled cutting. Aside from size and precision, the principal difference between these hobby-type machines and larger true vertical mills is that the X-Y table is at a fixed elevation; the Z-axis is controlled in basically the same fashion as drill press, where a larger vertical or knee mill has a vertically fixed milling head, and changes the X-Y table elevation. As well, a mill-drill often uses a standard drill press-type Jacob's chuck, rather than an internally tapered arbor that accepts collets. These are frequently of lower quality than other types of machines, but still fill the hobby role well because they tend to be benchtop machines with small footprints and modest price tags.

Horizontal Milling Machine

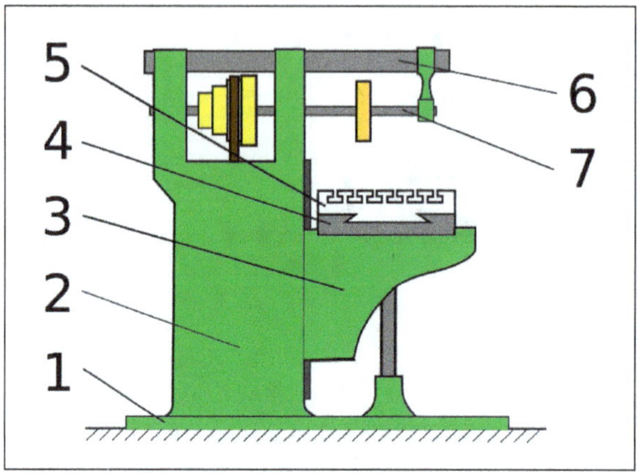

Horizontal milling machine. 1: base 2: column 3: knee 4 & 5: table (x-axis slide is integral) 6: overarm 7: arbor (attached to spindle).

A horizontal mill has the same sort but the cutters are mounted on a horizontal spindle across the table. Many horizontal mills also feature a built-in rotary table that allows milling at various angles; this feature is called a *universal table*. While endmills and the other types of tools available to a vertical mill may be used in a horizontal mill, their real advantage lies in arbor-mounted cutters, called side and face mills, which have a cross section rather like a circular saw, but are generally wider and smaller in diameter. Because the cutters have good support from the arbor and have a larger cross-sectional area than an end mill, quite heavy cuts can be taken enabling rapid material removal rates. These are used to mill grooves and slots. Plain mills are used to shape flat surfaces. Several cutters may be ganged together on the arbor to mill a complex shape of slots and planes. Special cutters can also cut grooves, bevels, radii, or indeed any section desired. These specialty cutters tend to be expensive. Simplex mills have one spindle, and duplex mills have two. It is also easier to cut gears on a horizontal mill. Some horizontal milling machines are equipped with a power-take-off provision on the table. This allows the table feed to be synchronized to a rotary fixture, enabling the milling of spiral features such as hypoid gears.

Comparative Merits

The choice between vertical and horizontal spindle orientation in milling machine design usually hinges on the shape and size of a workpiece and the number of sides of the workpiece that require machining. Work in which the spindle's axial movement is normal to one plane, with an endmill as the cutter, lends itself to a vertical mill, where the operator can stand before the machine and have easy access to the cutting action by looking down upon it. Thus vertical mills are most favored for diesinking work (machining a mould into a block of metal). Heavier and longer workpieces lend themselves to placement on the table of a horizontal mill.

Prior to numerical control, horizontal milling machines evolved first, because they evolved by putting milling tables under lathe-like headstocks. Vertical mills appeared in subsequent decades, and accessories in the form of add-on heads to change horizontal mills to vertical mills (and later vice versa) have been commonly used. Even in the CNC era, a heavy workpiece needing machining on multiple sides lends itself to a horizontal machining center, while diesinking lends itself to a vertical one.

Alternative Classifications

In addition to horizontal versus vertical, other distinctions are also important:

Criterion	Example classification scheme	Comments
Spindle axis orientation	Vertical versus horizontal; Turret versus non-turret.	Among vertical mills, "Bridgeport-style" is a whole class of mills inspired by the Bridgeport original, rather like the IBM PC spawned the industry of IBM-compatible PCs by other brands.
Control	Manual; Mechanically automated via cams; Digitally automated via NC/CNC.	In the CNC era, a very basic distinction is manual versus CNC. Among manual machines, a worthwhile distinction is non-DRO-equipped versus DRO-equipped.
Control (specifically among CNC machines)	Number of axes (e.g., 3-axis, 4-axis, or more).	Within this scheme, also: • Pallet-changing versus non-pallet-changing. • Full-auto tool-changing versus semi-auto or manual tool-changing.
Purpose	General-purpose versus special-purpose or single-purpose.	
Purpose	Toolroom machine versus production machine.	Overlaps with above.
Purpose	"Plain" versus "universal".	A distinction whose meaning evolved over decades as technology progressed, and overlaps with other purpose classifications above. Not relevant to today's CNC mills. Regarding manual mills, the common theme is that "plain" mills were production machines with fewer axes than "universal" mills; for example, whereas a plain mill had no indexing head and a non-rotating table, a universal mill would have those. Thus it was suited to universal service, that is, a wider range of possible toolpaths. Machine tool builders no longer use the "plain"-versus-"universal" labeling.
Size	Micro, mini, benchtop, standing on floor, large, very large, gigantic.	
Power source	Line-shaft-drive versus individual electric motor drive.	Most line-shaft-drive machines, ubiquitous circa 1880–1930, have been scrapped by now.
	Hand-crank-power versus electric.	Hand-cranked not used in industry but suitable for hobbyist micromills.

Variants

A Sieg X2 miniature hobbyist mill plainly showing the basic parts of a mill.

- Bed mill: This refers to any milling machine where the spindle is on a *pendant* that moves up and down to move the cutter into the work, while the table sits on a stout *bed* that rests on the floor. These are generally more rigid than a knee mill. Gantry mills can be included in this bed mill category.
- Box mill or column mill: Very basic hobbyist bench-mounted milling machines that feature a head riding up and down on a column or box way.
- C-frame mill: These are larger, industrial production mills. They feature a knee and fixed spindle head that is only mobile vertically. They are typically much more powerful than a turret mill, featuring a separate hydraulic motor for integral hydraulic power feeds in all directions, and a twenty to fifty horsepower motor. Backlash eliminators are almost always standard equipment. They use large NMTB 40 or 50 tooling. The tables on C-frame mills are usually 18" by 68" or larger, to allow multiple parts to be machined at the same time.
- Floor mill: These have a row of rotary tables, and a horizontal pendant spindle mounted on a set of tracks that runs parallel to the table row. These mills have predominantly been converted to CNC, but some can still be found (if one can even find a used machine available) under manual control. The spindle carriage moves to each individual table, performs the machining operations, and moves to the next table while the previous table is being set up for the next operation. Unlike other mills, floor mills have movable floor units. A crane drops massive rotary tables, X-Y tables, etc., into position for machining, allowing large and complex custom milling operations.
- Gantry mill: The milling head rides over two rails (often steel shafts) which lie at each side of the work surface.
- Horizontal boring mill: Large, accurate bed horizontal mills that incorporate many

features from various machine tools. They are predominantly used to create large manufacturing jigs, or to modify large, high precision parts. They have a spindle stroke of several (usually between four and six) feet, and many are equipped with a tailstock to perform very long boring operations without losing accuracy as the bore increases in depth. A typical bed has X and Y travel, and is between three and four feet square with a rotary table or a larger rectangle without a table. The pendant usually provides between four and eight feet of vertical movement. Some mills have a large (30" or more) integral facing head. Right angle rotary tables and vertical milling attachments are available for further flexibility.

- Jig borer, Vertical mills: That are built to bore holes, and very light slot or face milling. They are typically bed mills with a long spindle throw. The beds are more accurate, and the handwheels are graduated down to .0001" for precise hole placement.
- Knee mill or knee-and-column mill: Refers to any milling machine whose x-y table rides up and down the column on a vertically adjustable knee. This includes Bridgeports.
- Planer-style mill (Plano Milling): Large mills built in the same configuration as planers except with a milling spindle instead of a planing head. This term is growing dated as planers themselves are largely a thing of the past.
- Ram-type mill: This can refer to any mill that has a cutting head mounted on a sliding ram. The spindle can be oriented either vertically or horizontally. In practice most mills with rams also involve swiveling ability, whether or not it is called "turret" mounting. The Bridgeport configuration can be classified as a vertical-head ram-type mill. Van Norman specialized in ram-type mills through most of the 20th century. Since the wide dissemination of CNC machines, ram-type mills are still made in the Bridgeport configuration (with either manual or CNC control), but the less common variations (such as were built by Van Norman, Index, and others) have died out, their work being done now by either Bridgeport-form mills or machining centers.
- Turret mill: More commonly referred to as Bridgeport-type milling machines. The spindle can be aligned in many different positions for a very versatile, if somewhat less rigid machine.

Alternative Terminology

A milling machine is often called a mill by machinists. The archaic term miller was commonly used in the 19th and early 20th centuries.

Since the 1960s there has developed an overlap of usage between the terms milling machine and machining center. NC/CNC machining centers evolved from milling machines, which is why the terminology evolved gradually with considerable overlap that still persists. The distinction, when one is made, is that a machining center is a mill with features that pre-CNC mills never had, especially an automatic tool changer (ATC) that includes a tool magazine (carousel), and sometimes an automatic pallet changer (APC). In typical usage, all machining centers are mills, but not all mills are machining centers; only mills with ATCs are machining centers.

Computer Numerical Control

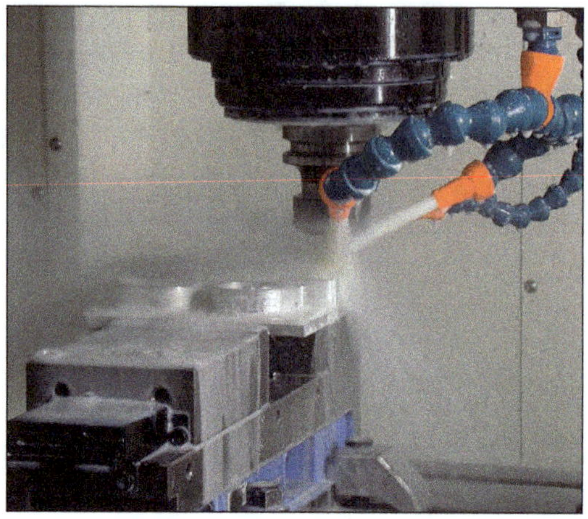

Thin wall milling of aluminum using a water based cutting fluid on the milling cutter.

Most CNC milling machines (also called *machining centers*) are computer controlled vertical mills with the ability to move the spindle vertically along the Z-axis. This extra degree of freedom permits their use in diesinking, engraving applications, and 2.5D surfaces such as relief sculptures. When combined with the use of conical tools or a ball nose cutter, it also significantly improves milling precision without impacting speed, providing a cost-efficient alternative to most flat-surface hand-engraving work.

Five-axis machining center with rotating table and computer interface.

CNC machines can exist in virtually any of the forms of manual machinery, like horizontal mills. The most advanced CNC milling-machines, the multiaxis machine, add two more axes in addition to the three normal axes (XYZ). Horizontal milling machines also have a C or Q axis, allowing the horizontally mounted workpiece to be rotated, essentially allowing asymmetric and eccentric turning. The fifth axis (B axis) controls the tilt of the tool itself. When all of these axes are used in conjunction with each other, extremely complicated geometries, even organic geometries such as a human head can be made with relative ease with these machines. But the skill to program such geometries is beyond that of most operators. Therefore, 5-axis milling machines are practically always programmed with CAM.

The operating system of such machines is a closed loop system and functions on feedback. These machines have developed from the basic NC (NUMERIC CONTROL) machines. A computerized form of NC machines is known as CNC machines. A set of instructions (called a program) is used to guide the machine for desired operations. Some very commonly used codes, which are used in the program are:

```
G00 - rapid traverse

G01 - linear interpolation of tool.

G21 - dimensions in metric units.

M03/M04 - spindle start (clockwise/counter clockwise).

T01 M06 - automatic tool change to tool 1

M30 - program end.
```

Various other codes are also used. A CNC machine is operated by a single operator called a programmer. This machine is capable of performing various operations automatically and economically.

With the declining price of computers and open source CNC software, the entry price of CNC machines has plummeted.

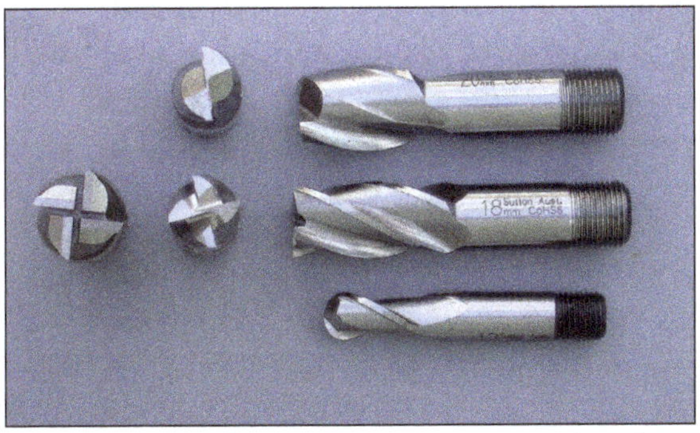

High speed steel with cobalt endmills used for cutting operations in a milling machine.

Tooling

The accessories and cutting tools used on machine tools (including milling machines) are referred to in aggregate by the mass noun "tooling". There is a high degree of standardization of the tooling used with CNC milling machines, and a lesser degree with manual milling machines. To ease up the organization of the tooling in CNC production many companies use a tool management solution.

Milling cutters for specific applications are held in various tooling configurations.

CNC milling machines nearly always use SK (or ISO), CAT, BT or HSK tooling. SK tooling is the most common in Europe, while CAT tooling, sometimes called V-Flange Tooling, is the oldest and probably most common type in the USA. CAT tooling was invented by Caterpillar Inc. of Peoria, Illinois, in order to standardize the tooling used on their machinery. CAT tooling comes in a range of sizes designated as CAT-30, CAT-40, CAT-50, etc. The number refers to the Association for Manufacturing Technology (formerly the National Machine Tool Builders Association (NMTB)) Taper size of the tool.

A CAT-40 toolholder.

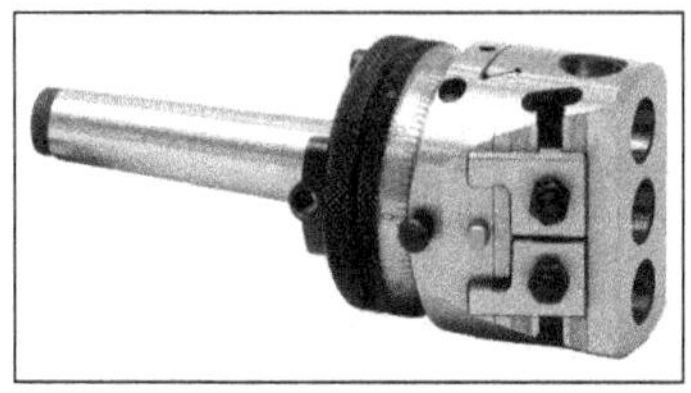

A boring head on a Morse taper shank.

An improvement on CAT Tooling is BT Tooling, which looks similar and can easily be confused with CAT tooling. Like CAT Tooling, BT Tooling comes in a range of sizes and uses the same NMTB body taper. However, BT tooling is symmetrical about the spindle axis, which CAT tooling is not. This gives BT tooling greater stability and balance at high speeds. One other subtle difference between these two toolholders is the thread used to hold the pull stud. CAT Tooling is all Imperial thread and BT Tooling is all Metric thread. Note that this affects the pull stud only; it does not affect the tool that they can hold. Both types of tooling are sold to accept both Imperial and metric sized tools.

SK and HSK tooling, sometimes called "Hollow Shank Tooling", is much more common in Europe where it was invented than it is in the United States. It is claimed that HSK tooling is even better than BT Tooling at high speeds. The holding mechanism for HSK tooling is placed within the (hollow) body of the tool and, as spindle speed increases, it expands, gripping the tool more tightly with increasing spindle speed. There is no pull stud with this type of tooling.

For manual milling machines, there is less standardization, because a greater plurality of formerly competing standards exist. Newer and larger manual machines usually use NMTB tooling. This tooling is somewhat similar to CAT tooling but requires a drawbar within the milling machine. Furthermore, there are a number of variations with NMTB tooling that make interchangeability troublesome. The older a machine, the greater the plurality of standards that may apply (e.g., Morse, Jarno, Brown & Sharpe, Van Norman, and other less common builder-specific tapers). However, two standards that have seen especially wide usage are the Morse #2 and the R8, whose prevalence was driven by the popularity of the mills built by Bridgeport Machines of Bridgeport, Connecticut. These mills so dominated the market for such a long time that "Bridgeport" is virtually synonymous with "manual milling machine". Most of the machines that Bridgeport made between 1938 and 1965 used a Morse taper #2, and from about 1965 onward most used an R8 taper.

Accessories

- Arbor support.
- Stop block.

CNC Pocket Milling

Pocket milling has been regarded as one of the most widely used operations in machining. It is extensively used in aerospace and shipyard industries. In pocket milling the material inside an arbitrarily closed boundary on a flat surface of a work piece is removed to a fixed depth. Generally flat bottom end mills are used for pocket milling. Firstly roughing operation is done to remove the bulk of material and then the pocket is finished by a finish end mill. Most of the industrial milling operations can be taken care of by 2.5 axis CNC milling. This type of path control can machine up to 80% of all mechanical parts. Since the importance of pocket milling is very relevant, therefore effective pocketing approaches can result in reduction in machining time and cost. NC pocket milling can be carried out mainly by two tool paths, viz. linear and non-linear.

Linear Tool Path

In this approach, the tool movement is unidirectional. Zig-zag and zig tool paths are the examples of linear tool path.

Zig-zag Tool Path

In zig-zag milling, material is removed both in forward and backward paths. In this case, cutting is done both with and against the rotation of the spindle. This reduces the machining time but increases machine chatter and tool wear.

Zig Tool Path

In zig milling, the tool moves only in one direction. The tool has to be lifted and retracted after each cut, due to which machining time increases. However, in case of zig milling surface quality is better.

Non-linear Tool Path

In this approach, tool movement is multi-directional. One example of non-linear tool path is contour-parallel tool path.

Contour-parallel Tool Path

In this approach, the required pocket boundary is used to derive the tool path. In this case the cutter is always in contact with the work material. Hence the idle time spent in positioning and retracting the tool is avoided. For large-scale material removal, contour-parallel tool path is widely used because it can be consistently used with up-cut or down-cut method during the entire process. There are three different approaches that fall into the category of contour-parallel tool path generation. They are:

- Pair-wise intersection approach:In pair-wise intersection approach, the boundary of the pocket is brought inwards in steps, The offset segments will intersect at concave corners. To obtain the required contour, these intersections are to be trimmed off. On the other hand, in case of convex corner, the offset segments are extended and thereby connected to make the contour. These operations viz. offsetting, trimming and extending are repeatedly done to cover the entire machining volume with sufficient layer of profiles.

- Voronoi diagram approach: In voronoi diagram approach, the pocket boundary is segmented and voronoi diagram is constructed for the entire pocket boundary. These voronoi diagrams are used for generating the tool path for machining. This method is considered to be more efficient and robust. Moreover, it avoids topological problems associated with traditional offsetting algorithms.

Curvilinear tool path

In this approach, the tool travels along a gradually evolving spiral path. The spiral starts at the center of the pocket to be machined and the tool gradually moves towards the pocket boundary. The direction of the tool path changes progressively and local acceleration and deceleration of the tool are minimized. This reduces tool wear.

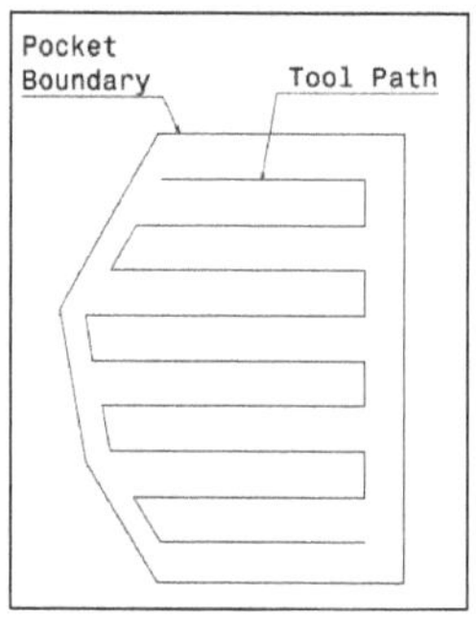

Zig-zag tool path.

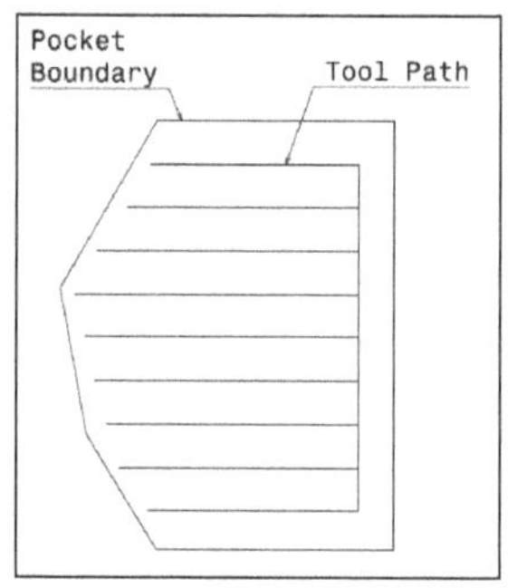

Zig tool path.

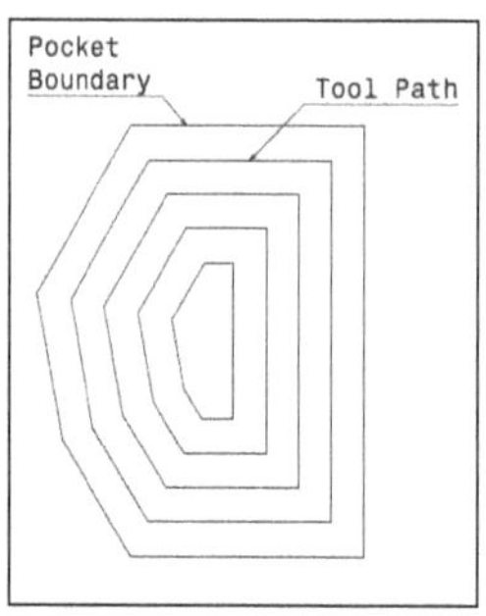

Contour-parallel tool path.

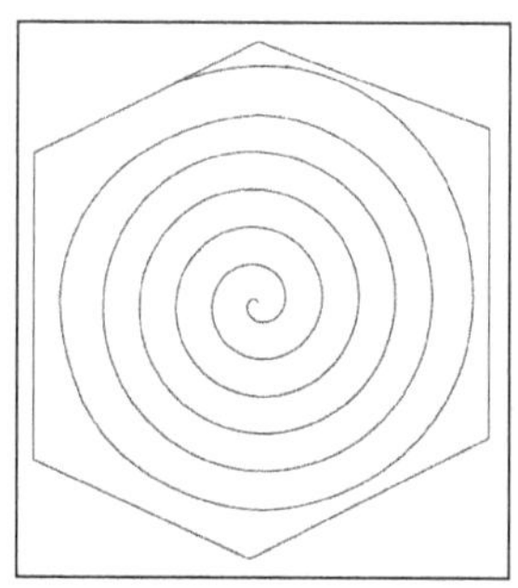

Curvilinear tool path.

TURNING

Turning is a machining process in which a cutting tool, typically a non-rotary tool bit, describes a helix toolpath by moving more or less linearly while the workpiece rotates.

Usually the term "turning" is reserved for the generation of *external* surfaces by this cutting action, whereas this same essential cutting action when applied to *internal* surfaces (that is, holes, of one kind or another) is called "boring". Thus the phrase "turning and boring" categorizes the larger family of (essentially similar) processes known as lathing. The cutting of faces on the workpiece (that is, surfaces perpendicular to its rotating axis), whether with a turning or boring tool, is called "facing", and may be lumped into either category as a subset.

Turning can be done manually, in a traditional form of lathe, which frequently requires continuous supervision by the operator, or by using an automated lathe which does not. Today the most common type of such automation is computer numerical control, better known as CNC. (CNC is also commonly used with many other types of machining besides turning.)

When turning, the workpiece (a piece of relatively rigid material such as wood, metal, plastic, or stone) is rotated and a cutting tool is traversed along 1, 2, or 3 axes of motion to produce precise diameters and depths. Turning can be either on the outside of the cylinder or on the inside (also known as boring) to produce tubular components to various geometries. Although now quite rare, early lathes could even be used to produce complex geometric figures, even the platonic solids; although since the advent of CNC it has become unusual to use non-computerized toolpath control for this purpose.

The turning processes are typically carried out on a lathe, considered to be the oldest machine tools, and can be of different types such as *straight turning*, *taper turning*, *profiling* or *external grooving*. Those types of turning processes can produce various shapes of materials such as *straight*, *conical*, *curved*, or *grooved* workpieces. In general, turning uses simple *single-point cutting* tools. Each group of workpiece materials has an optimum set of tools angles which have been developed through the years.

The bits of waste metal from turning operations are known as chips (North America), or swarf (Britain). In some areas they may be known as *turnings*.

The tool's axes of movement may be literally a straight line, or they may be along some set of curves or angles, but they are essentially linear (in the non mathematical sense).

A component that's is subject to turning operations can be termed as a "Turned Part" or "Machined Component". Turning operations are carried out on a lathe machine which can be manually or CNC operated.

Roughing, or rough turning.

Parting aluminium.

Finish turning.

Turning Operations

Turning specific operations include:

Turning

The general process of turning involves rotating a part while a single-point cutting tool is moved parallel to the axis of rotation. Turning can be done on the external surface of the part as well as the internal surface (the process known as boring). The starting material is generally a workpiece generated by other processes such as casting, forging, extrusion, or drawing.

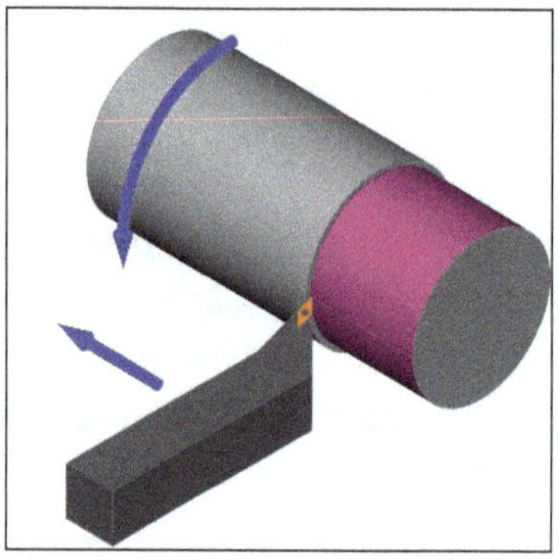

Turning.

Tapered Turning

Tapered turning produces a cylindrical shape that gradually decreases in diameter from one end to the other. This can be achieved a) from the compound slide b) from taper turning attachment c) using a hydraulic copy attachment d) using a C.N.C. lathe e) using a form tool f) by the offsetting of the tailstock - this method more suited for shallow tapers.

Spherical Generation

Spherical generation produces a spherical finished surface by turning a form around a fixed axis of revolution. Methods include a) using hydraulic copy attachment b) C.N.C. (computerised numerically controlled) lathe c) using a form tool (a rough and ready method) d) using bed jig (need drawing to explain).

Hard Turning

Hard turning is a type of turning done on materials with a Rockwell C hardness greater than 45. It is typically performed after the workpiece is heat treated.

The process is intended to replace or limit traditional grinding operations. Hard turning, when applied for purely stock removal purposes, competes favorably with rough grinding. However, when it is applied for finishing where form and dimension are critical, grinding is superior. Grinding produces higher dimensional accuracy of roundness and cylindricity. In addition, polished surface finishes of Rz=0.3-0.8z cannot be achieved with hard turning alone. Hard turning is appropriate for parts requiring a roundness accuracy of 0.5-12 micrometres, and/or surface roughness of Rz 0.8–7.0 micrometres. It is used for gears, injection pump components, and hydraulic components, among other applications.

Facing

Facing in the context of turning work involves moving the cutting tool at right angles to the axis of rotation of the rotating workpiece. This can be performed by the operation of the cross-slide, if one is fitted, as distinct from the longitudinal feed (turning). It is frequently the first operation performed in the production of the workpiece, and often the last—hence the phrase "ending up".

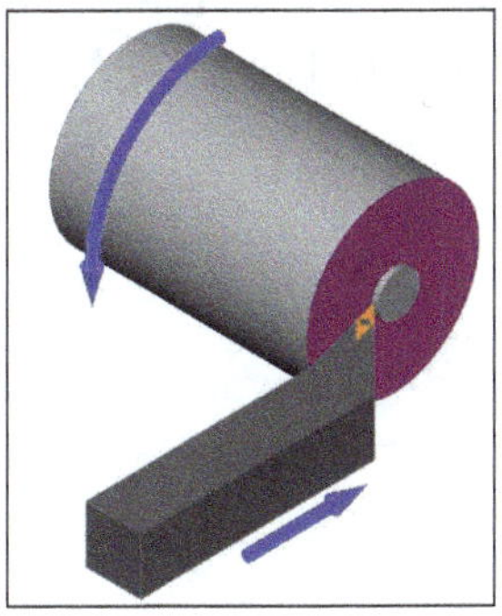

Facing.

Parting

This process, also called parting off or cutoff, is used to create deep grooves which will remove a completed or part-complete component from its parent stock.

Grooving

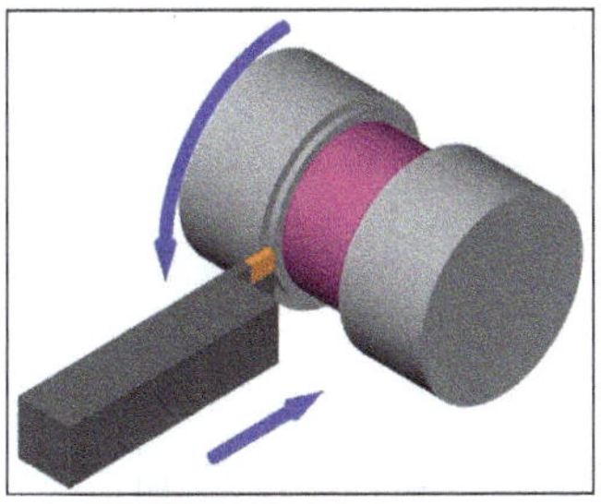

External grooving.

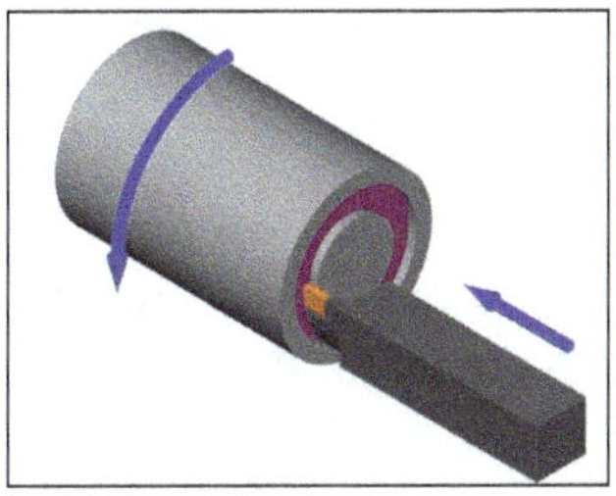

Face grooving.

Grooving is like parting, except that grooves are cut to a specific depth instead of severing a completed/part-complete component from the stock. Grooving can be performed on internal and external surfaces, as well as on the face of the part (face grooving or trepanning).

Non-specific operations include:

Boring

Enlarging or smoothing an existing hole created by drilling, moulding etc. i.e. the machining of internal cylindrical forms (generating) a) by mounting workpiece to the spindle via a chuck or faceplate b) by mounting workpiece onto the cross slide and placing cutting tool into the chuck. This work is suitable for castings that are too awkward to mount in the face plate. On long bed lathes large workpiece can be bolted to a fixture on the bed and a shaft passed between two lugs on the workpiece and these lugs can be bored out to size. A limited application but one that is available to the skilled turner/machinist.

Drilling

Is used to remove material from the inside of a workpiece. This process utilizes standard drill bits held stationary in the tail stock or tool turret of the lathe. The process can be done by separately available drilling machines.

Knurling.

Knurling

The cutting of a serrated pattern onto the surface of a part to use as a hand grip or as a visual enhancement using a special purpose knurling tool.

Reaming

The sizing operation that removes a small amount of metal from a hole already drilled. It is done for making internal holes of very accurate diameters. For example, a 6mm hole is made by drilling with 5.98 mm drill bit and then reamed to accurate dimensions.

Threading

Both standard and non-standard screw threads can be turned on a lathe using an appropriate cutting tool. (Usually having a 60, or 55° nose angle) Either externally, or within a bore. Generally referred to as single-point threading.

Tapping of threaded nuts and holes a) using hand taps and tailstock centre b)using a tapping device with a slipping clutch to reduce risk of breakage of the tap.

Threading operations include a)all types of external and internal thread forms using a single point tool also taper threads, double start threads, multi start threads, worms as used in worm wheel reduction boxes, leadscrew with single or multistart threads. b) by the use of threading boxes fitted with 4 form tools, up to 2" diameter threads but it is possible to find larger boxes than this.

Polygonal Turning

In which non-circular forms are machined without interrupting the rotation of the raw material.

Lathes

A lathe is a machine tool used principally for shaping pieces of metal, wood, or other materials by causing the workpiece to be held and rotated by the lathe while a tool bit is advanced into the work causing the cutting action. Lathes can be divided into three types for easy identification: engine lathe, turret lathe, and *special purpose lathes*. Some smaller ones are bench mounted and semi-portable. The larger lathes are floor mounted and may require special transportation if they must be moved. Field and maintenance shops generally use a lathe that can be adapted to many operations and that is not too large to be moved from one work site to another. The engine lathe is ideally suited for this purpose. A trained operator can accomplish more machining jobs with the engine lathe than with any other machine tool. Turret lathes and special purpose lathes are usually used in production or job shops for mass production or specialized parts, while basic engine lathes are usually used for any type of lathe work.

Workholding Methods

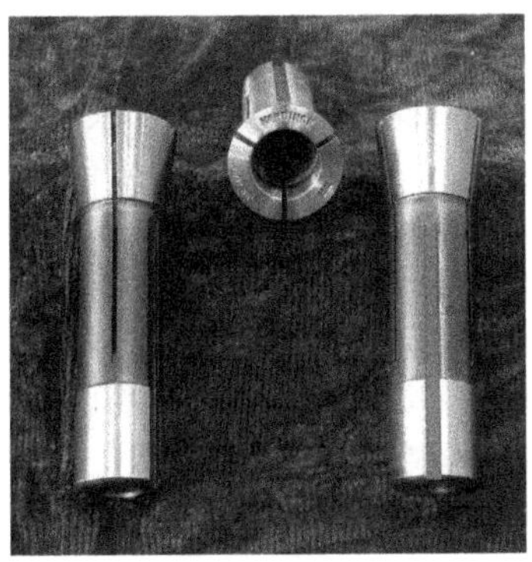

Collets.

- Chuck: Chucks are a very common workholding method. There are many types, some for round and square stock, and other for irregular shapes.
- Collet: Primarily used for small round workpieces.
- Faceplate: A faceplate, drive dog, and mandrel may be used to turn workpieces such as gear blanks.
- Centers: Pointed cones between which the part spins, often driven by a 'dog'. May be used to turn longer shafts and cylinders.
- Drive center: Use hydraulic or spring-loaded teeth that "bite" into the end of workpieces and can be used when the entire length of the workpiece must be machined.

Tooling

The various angles, shapes, and sizes of a *single-point cutting* tool have direct relation to the resulting surface of a workpiece in machining operations. Different types of angle such as *rake angle*, *side rake angle*, *cutting-edge angle*, *relief angle*, *nose radius* exist and may be different with respect to the workpiece. Also, there are many shapes of *single-point cutting* tools, such as *V-shaped* and *Square*. Usually, a special toolholder is used to hold the cutting tool firmly during operation.

Dynamics of Turning

Forces

The relative forces in a turning operation are important in the design of machine tools. The machine tool and its components must be able to withstand these forces without causing significant deflections, vibrations, or chatter during the operation. There are three principal forces during a turning process:

- The cutting or tangential force acts downward on the tool tip allowing deflection of the workpiece upward. It supplies the energy required for the cutting operation. The specific cutting force required to cut the material is called specific cutting force. Cutting force depends on the material.
- The axial or feed force acts in the longitudinal direction. It is also called the feed force because it is in the feed direction of the tool. This force tends to push the tool away from the chuck.
- The radial or thrust force acts in the radial direction and tends to push the tool away from the workpiece.

Speeds and Feeds

Speeds and feeds for turning are chosen based on cutter material, workpiece material, setup rigidity, machine tool rigidity and spindle power, coolant choice, and other factors.

Feed

- The distance the tool advances into the material in one revolution is called "feed". It is specified as mm per revolution (mm/rev).

DRILLING

Drilling is a cutting process that uses a drill bit to cut a hole of circular cross-section in solid materials. The drill bit is usually a rotary cutting tool, often multi-point. The bit is pressed against the work-piece and rotated at rates from hundreds to thousands of revolutions per minute. This forces the cutting edge against the work-piece, cutting off chips (swarf) from the hole as it is drilled.

In rock drilling, the hole is usually not made through a circular cutting motion, though the bit is usually rotated. Instead, the hole is usually made by hammering a drill bit into the hole with quickly repeated short movements. The hammering action can be performed from outside the hole (top-hammer drill) or within the hole (down-the-hole drill, DTH). Drills used for horizontal drilling are called drifter drills.

In rare cases, specially-shaped bits are used to cut holes of non-circular cross-section; a square cross-section is possible.

Titanium drilling.

Process

Drilled holes are characterized by their sharp edge on the entrance side and the presence of burrs on the exit side (unless they have been removed). Also, the inside of the hole usually has helical feed marks.

Drilling may affect the mechanical properties of the workpiece by creating low residual stresses around the hole opening and a very thin layer of highly stressed and disturbed material on the newly formed surface. This causes the workpiece to become more susceptible to corrosion and crack propagation at the stressed surface. A finish operation may be done to avoid these detrimental conditions.

For fluted drill bits, any chips are removed via the flutes. Chips may form long spirals or small flakes, depending on the material, and process parameters. The type of chips formed can be an indicator of the machinability of the material, with long chips suggesting good material machinability.

When possible drilled holes should be located perpendicular to the workpiece surface. This minimizes the drill bit's tendency to "walk", that is, to be deflected from the intended center-line of the bore, causing the hole to be misplaced. The higher the length-to-diameter ratio of the drill bit, the greater the tendency to walk. The tendency to walk is also preempted in various other ways, which include:

- Establishing a centering mark or feature before drilling, such as by:
    - Casting, molding, or forging a mark into the workpiece.
    - Center punching.
    - Spot drilling (i.e., center drilling).
    - Spot facing, which is machining a certain area on a casting or forging to establish an accurately located face on an otherwise rough surface.
- Constraining the position of the drill bit using a drill jig with drill bushings.

Surface finish produced by drilling may range from 32 to 500 microinches. Finish cuts will generate surfaces near 32 microinches, and roughing will be near 500 microinches.

Cutting fluid is commonly used to cool the drill bit, increase tool life, increase speeds and feeds, increase the surface finish, and aid in ejecting chips. Application of these fluids is usually done by flooding the workpiece with coolant and lubricant or by applying a spray mist.

In deciding which drill(s) to use it is important to consider the task at hand and evaluate which drill would best accomplish the task. There are a variety of drill styles that each serve a different purpose. The subland drill is capable of drilling more than one diameter. The spade drill is used to drill larger hole sizes. The indexable drill is useful in managing chips.

Spot Drilling

The purpose of spot drilling is to drill a hole that will act as a guide for drilling the final hole. The hole is only drilled part way into the workpiece because it is only used to guide the beginning of the next drilling process.

Center Drilling

Centre drill is A two-fluted tool consisting of a twist drill with a 60° countersink; used to drill countersink center holes in a workpiece to be mounted between centers for turning or grinding.

Deep Hole Drilling

Blast hole several meters long, drilled in granite.

Deep hole drilling is defined as drilling a hole of depth greater than ten times the diameter of the hole. These types of holes require special equipment to maintain the straightness and tolerances. Other considerations are roundness and surface finish.

Deep hole drilling is generally achievable with a few tooling methods, usually gun drilling or BTA drilling. These are differentiated due to the coolant entry method (internal or external) and chip removal method (internal or external). Using methods such as a rotating tool and counter-rotating workpiece are common techniques to achieve required straightness tolerances. Secondary tooling methods include trepanning, skiving and burnishing, pull boring, or bottle boring. Finally, a new kind of drilling technology is available to face this issue: vibration drilling. This technology breaks up the chips by a small controlled axial vibration of the drill. The small chips are easily removed by the flutes of the drill.

A high tech monitoring system is used to control force, torque, vibrations, and acoustic emission. Vibration is considered a major defect in deep hole drilling which can often cause the drill to break. A special coolant is usually used to aid in this type of drilling.

Gun Drilling

Gun drilling was originally developed to drill out gun barrels and is used commonly for drilling smaller diameter deep holes. The depth-to-diameter ratio can be even greater than 300:1. The key feature of gun drilling is that the bits are self-centering; this is what allows for such deep accurate holes. The bits use a rotary motion similar to a twist drill; however, the bits are designed with bearing pads that slide along the surface of the hole keeping the drill bit on center. Gun drilling is usually done at high speeds and low feed rates.

Trepanning

Trepanning is commonly used for creating larger diameter holes (up to 915 mm (36.0 in)) where a standard drill bit is not feasible or economical. Trepanning removes the desired diameter by cutting out a solid disk similar to the workings of a drafting compass. Trepanning is performed on flat products such as sheet metal, granite (curling stone), plates, or structural members like I-beams. Trepanning can also be useful to make grooves for inserting seals, such as O-rings.

Microdrilling

Microdrilling refers to the drilling of holes less than 0.5 mm (0.020 in). Drilling of holes at this small diameter presents greater problems since coolant fed drills cannot be used and high spindle speeds are required. High spindle speeds that exceed 10,000 RPM also require the use of balanced tool holders.

Vibration Drilling

The first studies into vibration drilling began in the 1950s (Pr. V.N. Poduraev, Moscow Bauman University). The main principle consists in generating axial vibrations or oscillations in addition to the feed movement of the drill so that the chips break up and are then easily removed from the cutting zone.

There are two main technologies of vibration drilling: self-maintained vibration systems and forced vibration systems. Most vibration drilling technologies are still at a research stage. In the case of self-maintained vibration drilling, the eigenfrequency of the tool is used in order to make it naturally vibrate while cutting; vibrations are self-maintained by a mass-spring system included in the tool holder. Other works use a piezoelectric system to generate and control the vibrations. These systems allow high vibration frequencies (up to 2 kHz) for small magnitude (about a few micrometers); they are particularly suitable for drilling small holes. Finally, vibrations can be generated by mechanical systems: the frequency is given by the combination of the rotation speed and the number of oscillation per rotation (a few oscillations per rotation), with magnitude about 0.1 mm.

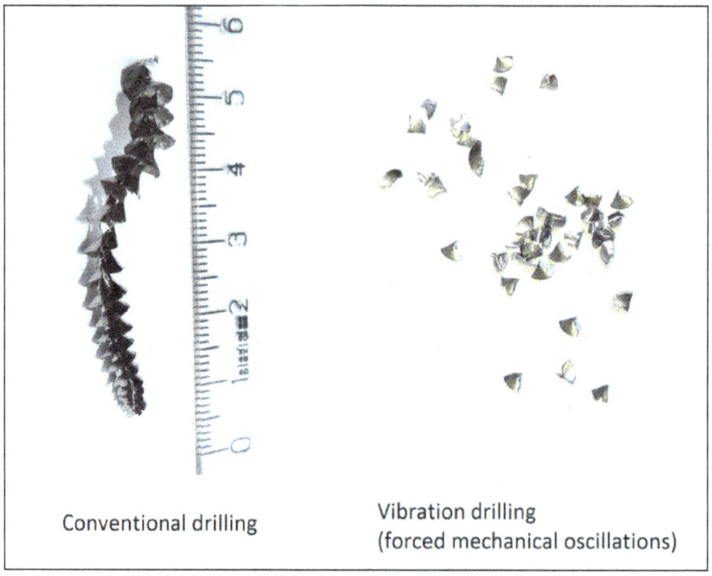

Titanium chips – conventional drilling vs vibration drilling.

Vibration drilling of an aluminum-CFRP multi-material stack with MITIS technology.

This last technology is a fully industrial one (example: SineHoling technology of MITIS). Vibration drilling is a preferred solution in situations like deep hole drilling, multi-material stack drilling (aeronautics) and dry drilling (without lubrication). Generally, it provides improved reliability and greater control of the drilling operation.

Circle Interpolating

Circle interpolating, also known as *orbital drilling,* is a process for creating holes using machine cutters.

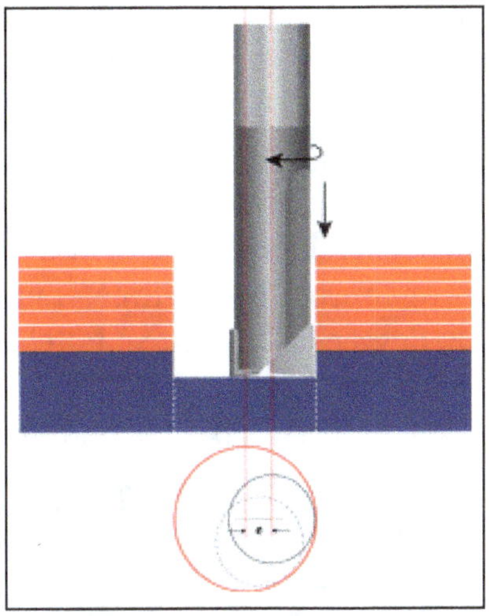

The orbital drilling principle.

Orbital drilling is based on rotating a cutting tool around its own axis and simultaneously about a centre axis which is off-set from the axis of the cutting tool. The cutting tool can then be moved simultaneously in an axial direction to drill or machine a hole – and/or combined with an arbitrary sidewards motion to machine an opening or cavity.

By adjusting the offset, a cutting tool of a specific diameter can be used to drill holes of different diameters as illustrated. This implies that the cutting tool inventory can be substantially reduced.

The term orbital drilling comes from that the cutting tool "orbits" around the hole center. The mechanically forced, dynamic offset in orbital drilling has several advantages compared to conventional drilling that drastically increases the hole precision. The lower thrust force results in a burr-less hole when drilling in metals. When drilling in composite materials the problem with delamination is eliminated.

Material

Drilling in Metal

High-speed steel twist bit drilling into aluminium with methylated spirits lubricant.

Under normal usage, swarf is carried up and away from the tip of the drill bit by the fluting of the drill bit. The cutting edges produce more chips which continue the movement of the chips outwards from the hole. This is successful until the chips pack too tightly, either because of deeper than normal holes or insufficient *backing off* (removing the drill slightly or totally from the hole while drilling). Cutting fluid is sometimes used to ease this problem and to prolong the tool's life by cooling and lubricating the tip and chip flow. Coolant may be introduced via holes through the drill shank, which is common when using a gun drill. When cutting aluminum in particular, cutting fluid helps ensure a smooth and accurate hole while preventing the metal from grabbing the drill bit in the process of drilling the hole. When cutting brass, and other soft metals that can grab the drill bit and causes "chatter", a face of approx. 1-2 millimeters can be ground on the cutting edge to create an obtuse angle of 91 to 93 degrees. This prevents "chatter" during which the drill tears rather than cuts the metal. However, with that shape of bit cutting edge, the drill is pushing the metal away, rather than grabbing the metal. This creates high friction and very hot swarf.

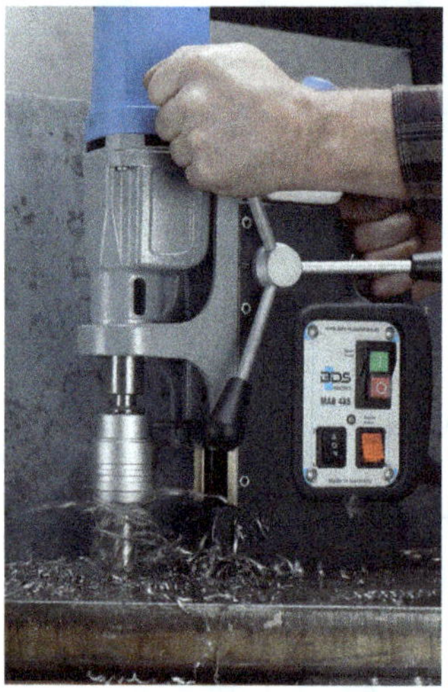

Magnetic Drilling Machine.

For heavy feeds and comparatively deep holes oil-hole drills are used in the drill bit, with a lubricant pumped to the drill head through a small hole in the bit and flowing out along the fluting. A conventional drill press arrangement can be used in oil-hole drilling, but it is more commonly seen in automatic drilling machinery in which it is the workpiece that rotates rather than the drill bit.

In computer numerical control (CNC) machine tools a process called *peck drilling*, or *interrupted cut drilling*, is used to keep swarf from detrimentally building up when drilling deep holes (approximately when the depth of the hole is three times greater than the drill diameter). Peck drilling involves plunging the drill part way through the workpiece, no more than five times the diameter of the drill, and then retracting it to the surface. This is repeated until the hole is finished. A modified form of this process, called *high speed peck drilling* or *chip breaking*, only retracts the drill slightly. This process is faster, but is only used in moderately long holes, otherwise it will overheat the drill bit. It is also used when drilling stringy material to break the chips.

When it is not possible to bring the material to the CNC machine, a Magnetic Base Drilling Machine may be used. The base allows drilling in a horizontal position and even on a ceiling. Usually, for these machines, it is better to use cutters because they can drill much faster with less speed. Cutter sizes vary from 12mm to 200mm DIA and from 30mm to 200mm DOC (depth of cut). These machines are widely used in construction, fabrication, marine, and oil & gas industries. In the oil and gas industry, pneumatic magnetic drilling machines are used to avoid sparks, as well as special tube magnetic drilling machines that can be fixed on pipes of different sizes, even inside.

Drilling in Wood

Wood being softer than most metals, drilling in wood is considerably easier and faster than drilling in metal. Cutting fluids are not used or needed. The main issue in drilling wood is ensuring clean entry and exit holes and preventing burning. Avoiding burning is a question of using sharp bits and the appropriate cutting speed. Drill bits can tear out chips of wood around the top and bottom of the hole and this is undesirable in fine woodworking applications.

The ubiquitous twist drill bits used in metalworking also work well in wood, but they tend to chip wood out at the entry and exit of the hole. In some cases, as in holes for rough carpentry, the quality of the hole does not matter, and a number of bits for fast cutting in wood exist, including spade bits and self-feeding auger bits. Many types of specialised drill bits for boring clean holes in wood have been developed, including brad-point bits, Forstner bits and hole saws. Chipping on exit can be minimized by using a piece of wood as backing behind the work piece, and the same technique is sometimes used to keep the hole entry neat.

Holes are easier to start in wood as the drill bit can be accurately positioned by pushing it into the wood and creating a dimple. The bit will thus have little tendency to wander.

Others

Some materials like plastics as well as other non-metals and some metals have a tendency to heat up enough to expand making the hole smaller than desired.

BROACHING

Broaching is a machining process that uses a toothed tool, called a broach, to remove material. There are two main types of broaching: linear and rotary. In linear broaching, which is the more common process, the broach is run linearly against a surface of the workpiece to effect the cut. Linear broaches are used in a broaching machine, which is also sometimes shortened to broach. In rotary broaching, the broach is rotated and pressed into the workpiece to cut an axisymmetric shape. A rotary broach is used in a lathe or screw machine. In both processes the cut is performed in one pass of the broach, which makes it very efficient.

Broaching is used when precision machining is required, especially for odd shapes. Commonly machined surfaces include circular and non-circular holes, splines, keyways, and flat surfaces. Typical workpieces include small to medium-sized castings, forgings, screw machine parts, and stampings. Even though broaches can be expensive, broaching is usually favored over other processes when used for high-quality production runs.

A push style $\frac{5}{16}$ inch (8 mm) keyway broach; note how the teeth are larger on the left end.

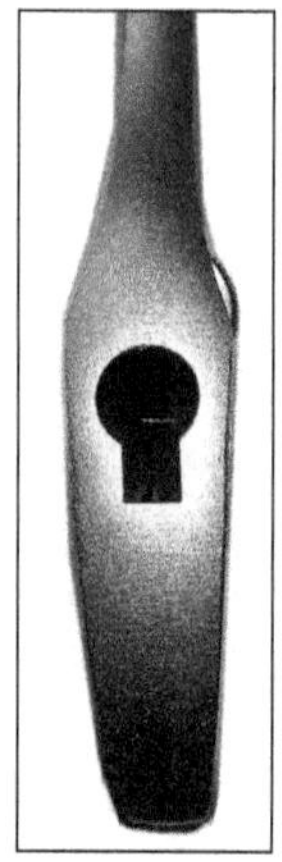

A broached keyway in the end of an adjustable wrench.

Broaches are shaped similar to a saw, except the height of the teeth increases over the length of the tool. Moreover, the broach contains three distinct sections: one for roughing, another for semi-finishing, and the final one for finishing. Broaching is an unusual machining process because it has the feed built into the tool. The profile of the machined surface is always the inverse of the profile of the broach. The rise per tooth (RPT), also known as the *step* or feed per tooth, determines the amount of material removed and the size of the chip. The broach can be moved relative to the workpiece or vice versa. Because all of the features are built into the broach, no complex motion or skilled labor is required to use it. A broach is effectively a collection of single-point cutting tools arrayed in sequence, cutting one after the other; its cut is analogous to multiple passes of a shaper.

Process

The process depends on the type of broaching being performed. Surface broaching is very simple as either the workpiece is moved against a stationary surface broach, or the workpiece is held stationary while the broach is moved against it.

Internal broaching is more involved. The process begins by clamping the workpiece into a special holding fixture, called a *workholder*, which mounts in the broaching machine. The broaching machine *elevator*, which is the part of the machine that moves the broach above the workholder, then lowers the broach through the workpiece. Once through, the broaching machine's *puller*, essentially a hook, grabs the *pilot* of the broach. The elevator then releases the top of the follower and the puller pulls the broach through the workpiece completely. The workpiece is then removed from the machine and the broach is raised back up to reengage with the elevator. The broach usually only moves linearly, but sometimes it is also rotated to create a spiral spline or gun-barrel rifling.

Cutting fluids are used for three reasons:

- To cool the workpiece and broach.
- To lubricate cutting surfaces.
- To flush the chips from the teeth.

Fortified petroleum cutting fluids are the most common. However, heavy-duty water-soluble cutting fluids are being used because of their superior cooling, cleanliness, and non-flammability.

Usage

An example of a broached workpiece. Here the broaching profile is a spline.

Broaching was originally developed for machining internal keyways. However, it was soon discovered that broaching is very useful for machining other surfaces and shapes for high volume workpieces. Because each broach is specialized to cut just one shape, either the broach must be specially designed for the geometry of the workpiece or the workpiece must be designed around a standard broach geometry. A customized broach is usually only viable with high volume workpieces, because the broach can cost US$15,000 to US$30,000 to produce.

Broaching speeds vary from 20 to 120 surface feet per minute (SFPM). This results in a complete cycle time of 5 to 30 seconds. Most of the time is consumed by the return stroke, broach handling, and workpiece loading and unloading.

The only limitations on broaching are that there are no obstructions over the length of the surface to be machined, the geometry to be cut does not have curves in multiple planes, and that the workpiece is strong enough to withstand the forces involved. Specifically for internal broaching a hole must first exist in the workpiece so the broach can enter. Also, there are limits on the size of internal cuts. Common internal holes can range from 0.125 to 6 in (3.2 to 152.4 mm) in diameter but it is possible to achieve a range of 0.05 to 13 in (1.3 to 330.2 mm). Surface broaches' range is usually 0.075 to 10 in (1.9 to 254.0 mm), although the feasible range is 0.02 to 20 in (0.51 to 508.00 mm).

Tolerances are usually ±0.002 in (±0.05 mm), but in precise applications a tolerance of ±0.0005 in (±0.01 mm) can be held. Surface finishes are usually between 16 and 63 microinches (μin), but can range from 8 to 125 μin. There may be minimal burrs on the exit side of the cut.

Broaching works best on softer materials, such as brass, bronze, copper alloys, aluminium, graphite, hard rubbers, wood, composites, and plastic. However, it still has a good machinability rating on mild steels and free machining steels. When broaching, the machinability rating is closely related to the hardness of the material. For steels the ideal hardness range is between 16 and 24 Rockwell C (HRC); a hardness greater than HRC 35 will dull the broach quickly. Broaching is more difficult on harder materials, stainless steel and titanium, but is still possible.

Types

Broaches can be categorized by many means:

- Use: Internal, or surface.

- Purpose: Single, or combination.
- Motion: Push, pull, or stationary.
- Construction: Solid, built-up, hollow or shell.
- Function: Roughing, sizing, or burnishing.

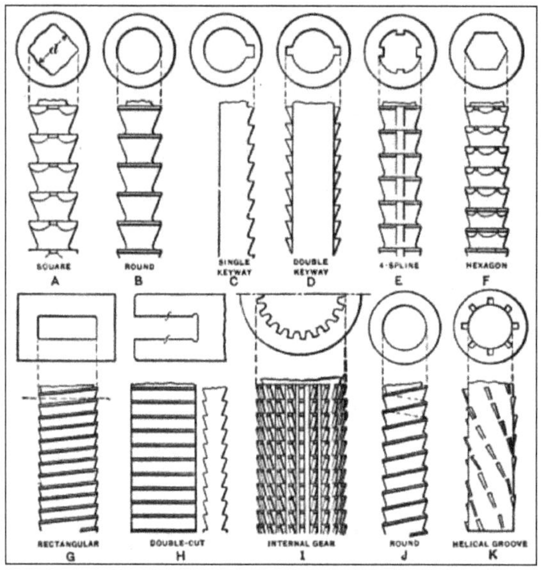

If the broach is large enough the costs can be reduced by using a *built-up* or *modular* construction. This involves producing the broach in pieces and assembling it. If any portion wears out only that section has to be replaced, instead of the entire broach.

Most broaches are made from high speed steel (HSS) or an alloy steel; Tin coatings are common on HSS to prolong life. Except when broaching cast iron, tungsten carbide is rarely used as a tooth material because the cutting edge will crack on the first pass.

Surface Broaches

The *slab broach* is the simplest surface broach. It is a general purpose tool for cutting flat surfaces.

Slot broaches (G & H) are for cutting slots of various dimensions at high production rates. Slot broaching is much quicker than milling when more than one slot needs to be machined, because multiple broaches can be run through the part at the same time on the same broaching machine.

Contour broaches are designed to cut concave, convex, cam, contoured, and irregular shaped surfaces.

Pot broaches are cut the inverse of an internal broach; they cut the outside diameter of a cylindrical workpiece. They are named after the pot looking fixture in which the broaches are mounted; the fixture is often referred to as a "pot". The pot is designed to hold multiple broaching tools concentrically over its entire length. The broach is held stationary while the workpiece is pushed or pulled through it. This has replaced hobbing for some involute gears and cutting external splines and slots.

Straddle broaches use two slab broaches to cut parallel surfaces on opposite sides of a workpiece in one pass. This type of broaching holds closer tolerances than if the two cuts were done independently. It is named after the fact that the broaches "straddle" the workpiece on multiple sides.

Internal Broaches

A modular broach.

Solid broaches are the most common type; they are made from one solid piece of material. For broaches that wear out quickly *shell* broaches are used; these broaches are similar to a solid broach, except there is a hole through the center where it mounts on an arbor. Shell broaches cost more initially, but save the cost overall if the broach must be replaced often because the pilots are on the mandrel and do not have to be reproduced with each replacement.

Modular broaches are commonly used for large internal broaching applications. They are similar to shell broaches in that they are a multi-piece construction. This design is used because it is cheaper to build and resharpen and is more flexible than a solid design.

A common type of internal broach is the *keyway* broach (C & D). It uses a special fixture called a *horn* to support the broach and properly locate the part with relation to the broach.

A *concentricity broach* is a special type of spline cutting broach which cuts both the minor diameter and the spline form to ensure precise concentricity.

The *cut-and-recut broach* is used to cut thin-walled workpieces. Thin-walled workpieces have a tendency to expand during cutting and then shrink afterward. This broach overcomes that problem by first broaching with the standard roughing teeth, followed by a "breathing" section, which serves as a pilot as the workpiece shrinks. The teeth after the "breathing" section then include roughing, semi-finishing, and finishing teeth.

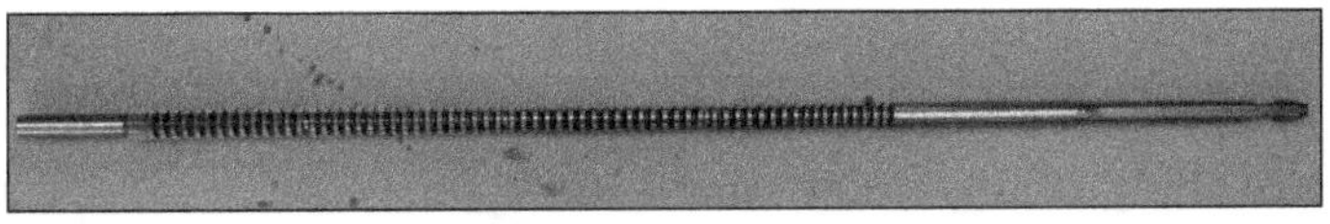

An internal broach for cutting splines.

The front pilot.

The finishing teeth.

The semi-finishing teeth.

The roughing teeth.

The slot in the tip of the broach where the broaching machine latches on to the broach to pull it through the workpiece.

Design

For defining the geometry of a broach an internal type is shown below. Note that the geometries of other broaches are similar.

The most important characteristic of a broach is the rise per tooth (RPT), which is how much material is removed by each tooth. The RPT varies for each section of the broach, which are the roughing section (t_r), semi-finishing section (t_s), and finishing section (t_f). The roughing teeth remove most of the material so the number of roughing teeth required dictates how long the broach is. The semi-finishing teeth provide surface finish and the finishing teeth provide the final finishing. The finishing section's RPT (t_f) is usually zero so that as the first finishing teeth wear the later ones continue the sizing function. For free-machining steels the RPT ranges from 0.006 to 0.001 in (0.152 to 0.025 mm). For surface broaching the RPT is usually between 0.003 to 0.006 in (0.076 to 0.152 mm) and for diameter broaching is usually between 0.0012 to 0.0025 in (0.030 to 0.064 mm). The exact value depends on many factors. If the cut is too big it will impart too much stress into the teeth and the workpiece; if the cut is too small the teeth rub instead of cutting. One way to increase the RPT while keeping the stresses down is with *chip breakers*. They are notches in the teeth designed to break the chip and decrease the overall amount of material being removed by any given tooth. For broaching to be effective, the workpiece should have 0.020 to 0.025 in (0.51 to 0.64 mm) more material than the final dimension of the cut.

The *hook* (α) angle is a parameter of the material being cut. For steel, it is between 15 and 20° and for cast iron it is between 6 and 8°. The *back-off* (γ) provides clearance for the teeth so that they don't rub on the workpiece; it is usually between 1 and 3°.

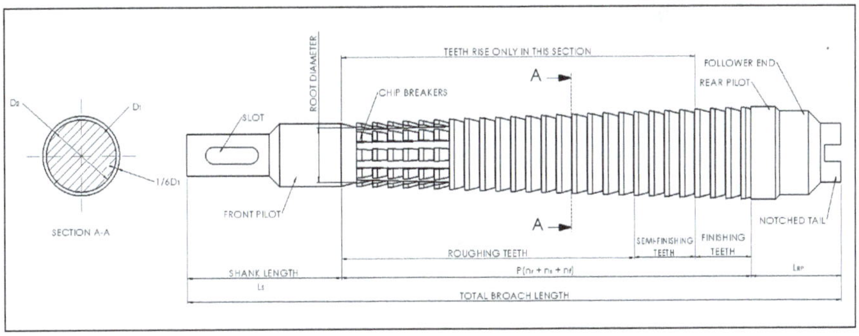

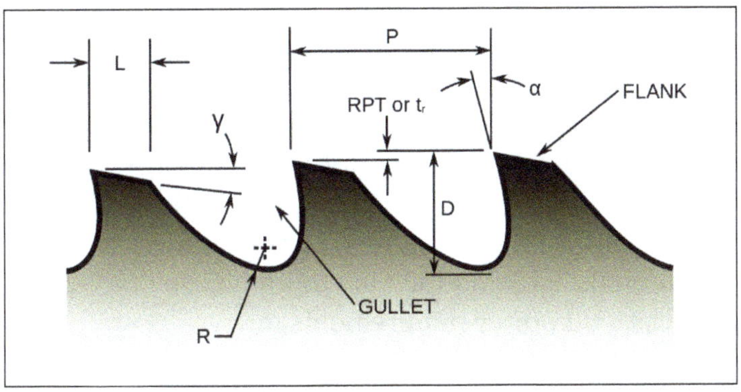

where:

- P = pitch.
- RPT = rise per tooth.
- n_r = number of roughing teeth.
- n_s = number of semi-finishing teeth.
- n_f = number of finishing teeth.
- t_r = RPT for the roughing teeth.
- t_s = RPT for the semi-finishing teeth.
- t_f = RPT for the finishing teeth.
- L_s = Shank length.
- L_{RP} = Rear pilot length.
- D_1 = Diameter of the tooth tip.
- D_2 = Diameter of the tooth root.
- D = Depth of a tooth (0.4P).
- L = Land (behind the cutting edge) (0.25P).
- R = Radius of the gullet (0.25P).
- α = Hook angle or rake angle.

- γ = Back-off angle or clearance angle.
- L_w = Length of the workpiece.

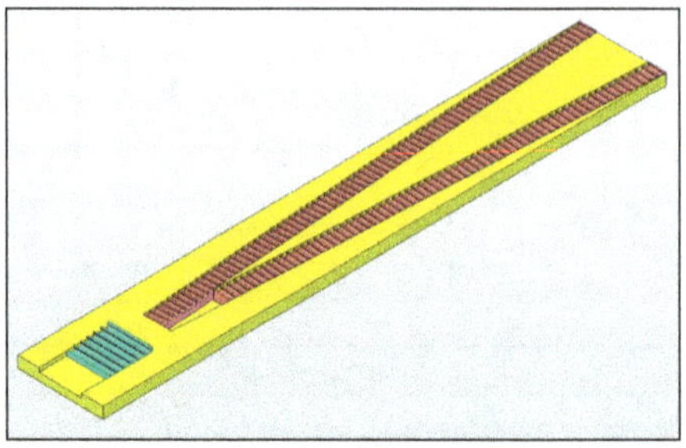

A progressive surface broach.

When radially broaching workpieces that require a deep cut per tooth, such as forgings or castings, a *rotor-cut* or *jump-cut* design can be used; these broaches are also known as *free egress* or *nibbling* broaches. In this design the RPT is designated to two or three rows of teeth. For the broach to work the first tooth of that cluster has a wide notch, or undercut, and then the next tooth has a smaller notch (in a three tooth design) and the final tooth has no notch. This allows for a deep cut while keeping stresses, forces, and power requirements low.

There are two different options for achieving the same goal when broaching a flat surface. The first is similar to the rotor-cut design, which is known as a *double-cut* design. Here four teeth in a row have the same RPT, but each progressive tooth takes only a portion of the cut due to notches in the teeth. The other option is known as a *progressive* broach, which completely machines the center of the workpiece and then the rest of the broach machines outward from there. All of these designs require a broach that is longer than if a standard design were used.

For some circular broaches, *burnishing teeth* are provided instead of finishing teeth. They are not really teeth, as they are just rounded discs that are 0.001 to 0.003 in (0.025 to 0.076 mm) oversized. This results in burnishing the hole to the proper size. This is primarily used on non-ferrous and cast iron workpieces.

The pitch defines the tooth construction, strength, and number of teeth in contact with the workpiece. The pitch is usually calculated from workpiece length, so that the broach can be designed to have at least two teeth in contact with the workpiece at any time; the pitch remains constant for all teeth of the broach. One way to calculate the pitch is:

$$P \cong 0.35\sqrt{L_w}$$

Example of a double-cut surface broach.

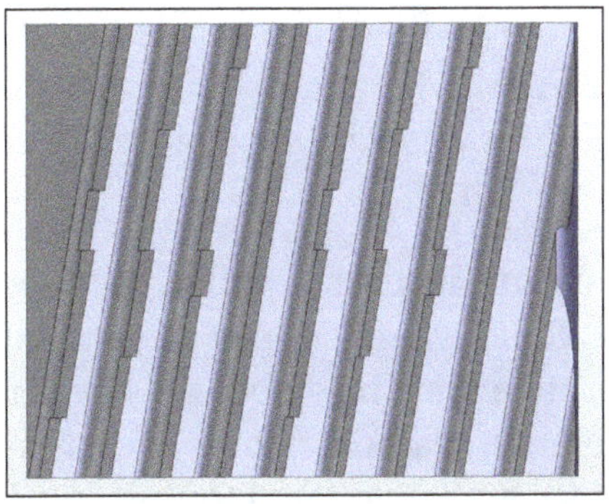

Top view of a double-cut surface broach.

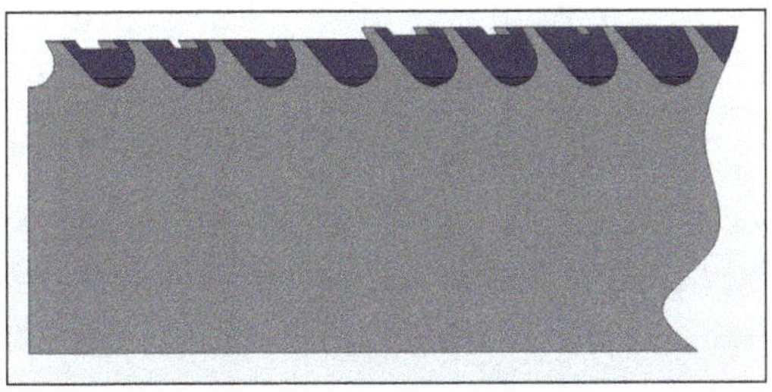

Side view of a double-cut surface broach.

Broaching Machines

Broaching machines are relatively simple as they only have to move the broach in a linear motion at a predetermined speed and provide a means for handling the broach automatically. Most machines are hydraulic, but a few specialty machines are mechanically driven. The machines are distinguished by whether their motion is horizontal or vertical. The choice of machine is primarily dictated by the stroke required. Vertical broaching machines rarely have a stroke longer than 60 in (1.5 m).

The hydraulic cylinder of a horizontal broaching machine.

Vertical broaching machines can be designed for push broaching, pull-down broaching, pull-up broaching, or surface broaching. Push broaching machines are similar to an arbor press with a guided ram; typical capacities are 5 to 50 tons. The two ram pull-down machine is the most common type of broaching machine. This style machine has the rams under the table. Pull-up machines have the ram above the table; they usually have more than one ram. Most surface broaching is done on a vertical machine.

Horizontal broaching machines are designed for pull broaching, surface broaching, continuous broaching, and rotary broaching. Pull style machines are basically vertical machines laid on the side with a longer stroke. Surface style machines hold the broach stationary while the workpieces are clamped into fixtures that are mounted on a conveyor system. Continuous style machines are similar to the surface style machines except adapted for internal broaching.

Horizontal machines used to be much more common than vertical machines; however, today they represent just 10% of all broaching machines purchased. Vertical machines are more popular because they take up less space.

Broaching is often impossible without the specific broaching or keyway machines unless you have a system that can be used in conjunction with a modern machining centre or driven tooling lathe; these extra bits of equipment open up the possibility of producing keyways, splines and torx through one-hit machining.

Rotary Broaching

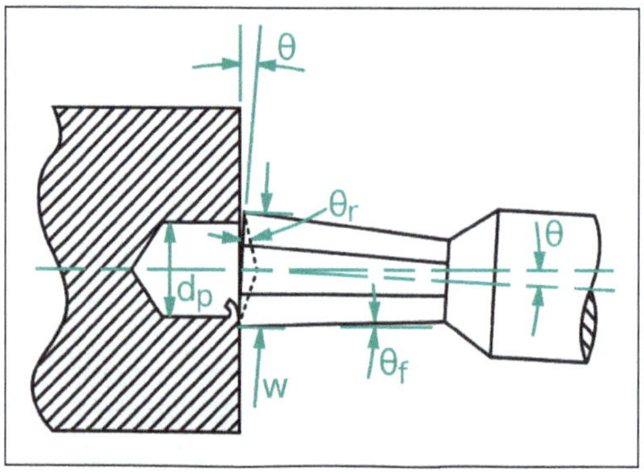

Schematic of a rotary broach starting a cut.

θ	Off-axis (wobble) angle
θ_r	Rake
θ_f	Front relief
d_p	Pilot diameter
w	Width across corners (AC)

A somewhat different design of cutting tool that can achieve the irregular hole or outer profile of a broach is called a *rotary broach* or *wobble broach*. One of the biggest advantages to this type of broaching is that it does not require a broaching machine, but instead is used on lathes, milling machines, screw machines or Swiss lathes.

Rotary broaching requires two tooling components: a tool holder and a broach. The leading (cutting) edge of the broach has a contour matching the desired final shape. The broach is mounted in a special tool holder that allows it to freely rotate. The tool holder is special because it holds the tool so that its axis of rotation is inclined slightly to the axis of rotation of the work. A typical value for this misalignment is 1°. This angle is what produces a rotating edge for the broach to cut the workpiece. Either the workpiece or the tool holder is rotated. If the tool holder is rotated, the misalignment causes the broach to appear as though it is "wobbling", which is the origin of the term *wobble broach*.

For internal broaching the sides of the broach are drafted inward so it becomes thinner; for external broaching the sides are drafted outward, to make the pocket bigger. This draft keeps the broach from jamming; the draft must be larger than the angle of misalignment. If the work piece rotates, the broach is pressed against it, is driven by it, and rotates synchronously with it. If the tool holder rotates, the broach is pressed against the workpiece, but is driven by the tool holder.

Ideally the tool advances at the same rate that it cuts. The ideal rate of cut is defined as:

Rate of cut [inches per rotation (IPR)] = (diameter of tool [inches]) × sin(Angle of misalignment [degrees])

If it advances much faster, then the tool becomes choked; conversely, if it advances much slower, then an interrupted or zig-zag cut occurs. In practice the rate of cut is slightly less than the ideal rate so that the load is released on the non-cutting edge of the tool.

There is some spiraling of the tool as it cuts, so the form at the bottom of the workpiece may be rotated with respect to the form at the top of the hole or profile. Spiraling may be undesirable because it binds the body of the tool and prevents it from cutting sharply. One solution to this is to reverse the rotation in mid cut, causing the tool to spiral in the opposite direction. If reversing the machine is not practical, then interrupting the cut is another possible solution.

In general, a rotary broach will not cut as accurately as a push or pull broach. However, the ability to use this type of cutting tool on common machine tools is highly advantageous.

MASS FINISHING

Mass finishing is the general description for vibrating or flowing an abrasive media around usually a number of non-fixtured parts, moving randomly within the mass of the abrasive media.

Various types of equipment generate energy that is transferred thru the media to the part being processed. The transfer of energy and randomly moving parts automates the finishing process, with part loading and unloading to be addressed.

There are a number of variables that affect mass finishing which includes:

- The equipment,
- Media,
- Soap compounds,
- The process.

The effects mass finishing has on parts include:

- Deburring,
- Refining RMS finishes,
- Burnishing (Brighting),
- Improving strength,
- Pre plate,
- Pre paint adhesion,
- Improving oil retention on surfaces,
- Cleaning.

Mass finishing processes are repeatable and one of the least expensive ways to automate finishes on parts.

Mass Finishing Machine Types

Tub Vibratory Machines

The tub vibratory machine is a horizontal urethane lined container with a u-shaped bottom and an open top. It can be manufactured with a variety of channel diameters and lengths making it an excellent choice for long or large parts. The tub can also be divided to run large parts preventing them from touching one another. The tub, built with extended lengths, lends itself to straight-line automation of vibrated parts.

The tub vibratory machine generates quite a bit more energy than rotating barrels and a minor amount of energy faster than bowl vibratory machines. The system is commonly powered or vibrated by a shaft with an eccentric attached to the tub. The shaft is rotated by a v belt, sheave and motor drive. The eccentric weights can be added or subtracted and adjusted around the shaft to increase or decrease vibration. There is another uncommon tub vibratory drive system that is driven with magnets. The tub vibratory machines were developed in the mid 1920s and were a welcome improvement over the rotating barrels that were in production prior to the tubs inception.

Bowl Vibratory Machine

The bowl vibratory machine is a horizontal urethane lined donut shaped container with an open top and vertical eccentrics shaft drive built thru the center column. The media mass rotates around the center column in a rolling forward feed motion. The machine can be built with a variety of channels and overall diameters; it is now being built with multiple channels feeding parts from one channel to another, allowing a variety of processes and drying with one machine.

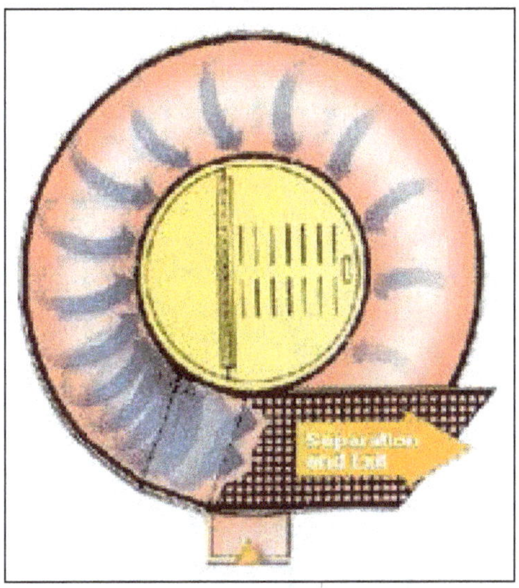

The bowl vibratory machine tends to be a better finishing machine than the tub; it does a better job of keeping parts relatively orientated to minimize damage than a tub. The bowl can also be divided for large parts.

The most significant advantage of a bowl over a tub is that ramps, gates and screens can be added to the bowl allowing self unload of the parts away from the media and out of the machine.

The self-unload and improved finishing process of the bowl makes it a more popular choice over the tub.

The bowl vibratory machines were developed in the 1950s with self-unloading versions being patented in the 60s. The bowl vibratory machines were again a welcomed improvement over the tub and barrel systems previous to it.

Centrifugal Disc

The centrifugal disc is a stationary urethane lined bowl with a rotating bottom urethane lined disc that accelerates the media outward to the sidewall which then decelerates and moves inward to the center of the disc for reacceleration. This change in acceleration and deceleration of the media creates a flowing action over the part, generating energy up to 15 times faster than vibratory with the increased energy. The disc machine will deburr mall tough parts and achieve very low RMS finishes on parts for pre plate or pre electro polished finishes.

The disc was developed in the late 60s and is being utilized today because it's easy to load and unload. Decreased process time makes it highly favorable for cellular manufacturing.

Centrifugal Barrel

The centrifugal barrel machine has enclosed urethane-lined barrels mounted on a turret. As the turret rotates one direction the barrels are belt or gear driven to rotate to a given ratio in the opposite direction. The individual barrels are loaded with the media, parts to be processed, water and compounds. The action within the barrel is a compressive sliding action creating the maximum energy available today in mass finishing.

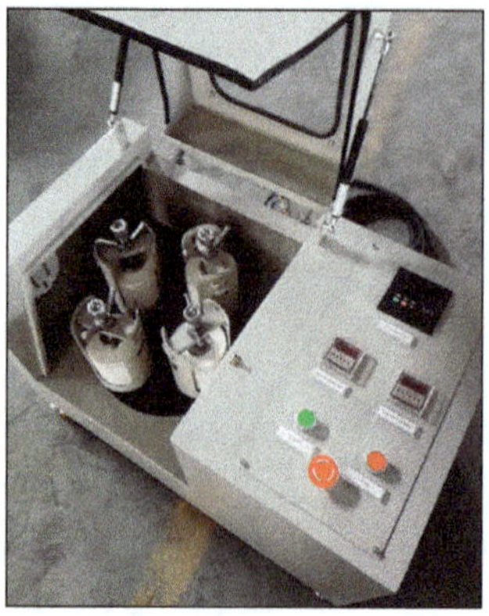

The process is excellent for small parts with large burrs. This system has the energy to drive small media into hard to get to areas. The system can also utilize one media to cut and brighten with one step. The bowl can also be divided for large parts.

The centrifugal barrel is also being used to brighten parts with dry medias to get as close to an as buffed finish as possible.

The centrifugal barrel machine was developed in the late 60s and is becoming more and more popular as processes are being developed. One of the disadvantages of the centrifugal barrel finishing is the load and unload of parts is a bit labor intensive.

Spindle Machines

The spindle machine is a variable speed rotating bowl of random shaped or preformed media. The spinning of the bowl slings the media to the outside wall. The part is rotated by a spindle that holds it into the spinning mass. The collision of the media hitting the part as it rotates by is what creates the energy.

The individual part holding spindle can be programmed to rotate at various speeds, and can orientate the part to be worked in certain areas or dwelled to concentrate in an area. The flexibility of orientation, rotation and dwell of the spindle allows parts such as gears to be worked to achieve different radius in various areas of the gear teeth.

Spindle machines produce higher energy levels than vibratory machines. Spindles were developed in the 70s and have been popular for gear deburring. The American-built spindle machine generally uses 2 to 4 spindles per machine. Other manufacturers tend to utilize many more spindles in a given machine.

Dry process spindle machines are used to obtain an as buffed finish on large complex parts such as band instruments.

Drag Machines

Drag machines are bowl machines that utilize light vibration with fixtures holding and dragging the part thru a heavy mass.

The drag machine produces more work than conventional vibratory systems resulting in reduced time cycles on complex parts.

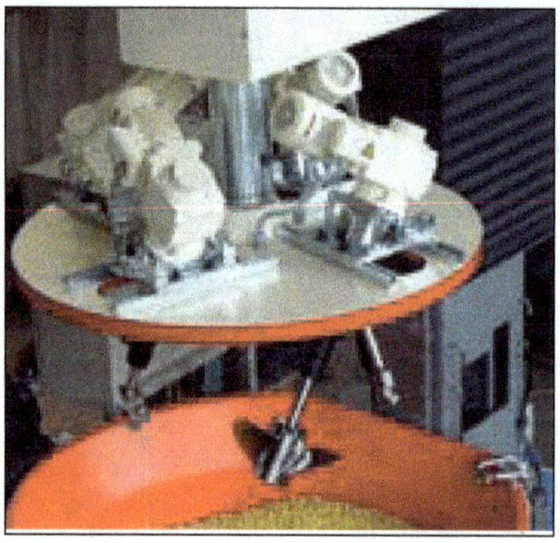

These systems have been a recent development and are used on a variety of parts like boat propellers, and air tool handles.

Barrel Finishing

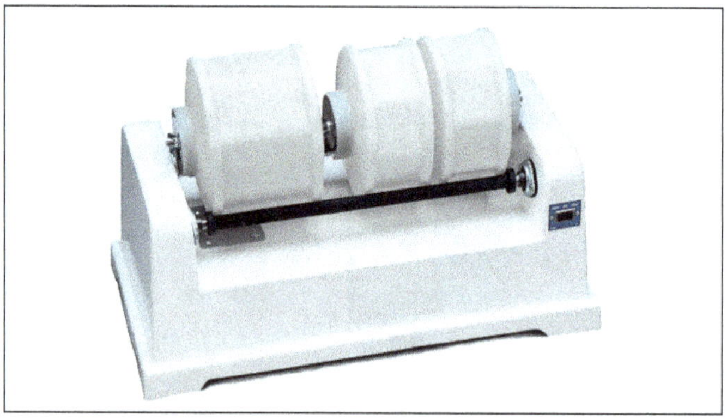

A hexagon or octagon barrel rotating horizontal will process parts slower than vibratory finishing, but produces amazing finishing results.

The barrel is rubber or urethane lined to protect parts. As the barrel rotates it lifts the media that eventually slides. The slide of the media containing the parts is the working action of the barrel. A faster RPM results in a thinner slide area, a slower RPM produces a thicker slid, 18-22 RPMS tends to be the optional slide for most applications.

The process involves water and compound, the more water and compound used the more cushioning action is achieved. Normal water levels are just above the media & parts. Media level is very important, 50% full with parts will produce the longest slide and therefore is a good starting point. Higher media levels will shorten the slide and provide more part protection.

Time cycles range from 1 to 24 hours, 1 for burnishing, 2 for deburring and longer cycles when deburring and burnishing in on operation.

The load and unload of barrels is labor intensive; therefore vibratory finishing has replaced many barrel applications. Internal areas of parts are not finished well with barrel processes, but when it comes to keeping flat parts from sticking together or super luster finishes without part damage, barrels have always been an excellent choice.

Media Types

Vibratory or mass finishing media transmits the energy from the machine to the work piece. It carries the compound for cleaning or brightening and it suspends and protects the parts.

The weight and size of the media has a lot to do with the amount of work it is capable of doing. Heavier and larger media cut faster and do more work because of the heavier hammer effect. The larger medias also create more voids within the mass allowing a more aggressive action. A smaller media will create a tighter mass capable of holding more fluid, and with less voids within the mass it will work to protect the part.

The shape of the media allows it to get into an area to get something done or stay out of an area so it will not lodge. A round or cylindrical shape will finish better because of single point contact with the part, like a ball pean hammer effect. A flat shape, such as a triangle, has a wide surface to rub over an edge allowing quicker deburring.

Media hardness has a lot to do with the finish depending on the type of material and hardness of the part. This and other factors will be discussed under media selection.

Media Type	Color	Approx. Wt./Cu.ft.	Cutting Ability	Surface Finish	Media Hardness	General Comments
Ceramic						All ceramics are cleaner running and in general, less expensive to operate than Plastic Medias. They give a scratchy Matte. Finish Best RMS Finish in vibratory is app. 12.
High Density	Brown	120 lbs.	Fastest	Dull Rough Matte Finish	60-65 Brinell	
Super Fast	Grey	70-90 lbs.	Very Fast	Scratchy Matte Finish	60-65 Brinell	
Fast Cut	Grey	70-90 lbs.	Fast	Scratchy Matte Finish	60-65 Brinell	
Medium Cut	Grey	70-90 lbs.	Medium Fast	Matte Finish	60-65 Brinell	
Light Weight	White	40 lbs.	Very Fast	Matte Finish	50 Brinell	
Polish	White	70-90 lbs.	Very Little	Bright Reflect	70-90 Brinell	
Plastics Synthetic: Plastic						Very clean unning media. All medias are biodegradable. All synthetics are capable of a preplate finish depending on metal type. This media will cut a burr when a ceramic media will roll it over.
Fast	Pink	45 lbs.	Fastest of Synthetics	Good (Matte)	0 on Brinell	
Medium	Tangerine	45 lbs.		Good (Matte)	0 on Brinell	
Preplate	Light-Green	45 lbs.		Bright Matte	0 on Brinell	
SP	Tan	45 lbs.	Slowest	Bright Polish	0 on Brinell	
Polyester: Plastic						The media =ardness is 45 Brinell. Longer wearing than the synthetic medias.
High Density	Grey	55 lbs.	Fastest of Polyester	Good(Dull Matte)	42-45 Brinell	
Fast & Medium	Green/ Red	55 lbs.		Very Good(Dull)	42-45 Brinell	
Preplate	Light Green	55 lbs.	Slowest	Dull Matte	42-45 Brinell	

Ceramic Media

Ceramic media is a very popular choice in finishing processes because of its long lasting and clean running characteristics. There are different grades of cut within the ceramic media family:

- Polish,
- Super Fast Cut,
- Medium Cut,
- High Density Fast Cut,
- Fast Cut,
- Light Weight (low density).

Below are the factors affecting media cutting performance:

Abrasive Size And Percentage - Polish medias have no abrasives where faster cut medias have courser grits and more of it.

Bonding Agent Strength - Faster cutting medias are bonded softer allowing the bond to wear away exposing new abrasives which increases cut rates. Polish medias are very hard.

The Ceramic Itself - Different clays will assist cut and affect weights of medias. Heavier and larger media will cut faster.

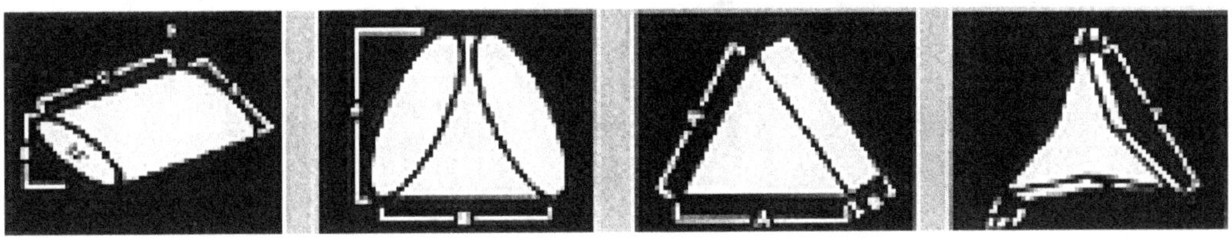

The manufacturing process determines the shapes ceramic medias are available in. The ceramic and abrasive are mixed wet or green and then extruded thru dyes. This limits the shapes available which are equal sides. As the media is extruded it is cut for thickness and then baked in a continuous or batch klin to dry and harden. The ceramic cone and ball are a couple exceptions to the extruded equal side shapes.

Ceramic is a good all around media and is one of the first choices to make for all metals. The 45 to 65 rockwell hardness of ceramic can sometimes damage softer metal, testing prior to processing is always advised.

Plastic Media

Plastic media is known by its bonding agent. The two primary bonding agents of plastic are:

- Polyester: Polyester is the original formulation of plastic media. It is produced from an oil derivative, therefore, non-biodegradable. The polyester breakdown is a white chalky sediment that's difficult to clean off of parts.
- Synthetic: Synthetic plastic is a recent development and runs cleaner than polyester. Synthetic is softer and lighter than polyester achieving superior finishes. Synthetic is also biodegradable.

The cutting element in plastic medias is usually quartz. The cutting element is limited to a given percentage because the manufacturer's process starts with a liquid and like adding sugar to water and mixing it, if too much sugar is added it will drop out. After the quartz is added a catalyst is added to harden the mix that's poured in a mold. The molds provide a variety of interesting shapes: cones, pyramid, tetrahedron and wedges.

Plastic medias are processed in a variety of different cuts:

- Fine Cut (Pre Plate),
- Medium Cut,

- Fast Cut,
- Extra Fast Cut (High Density Zirconium, 100 LBS Cu. Ft).

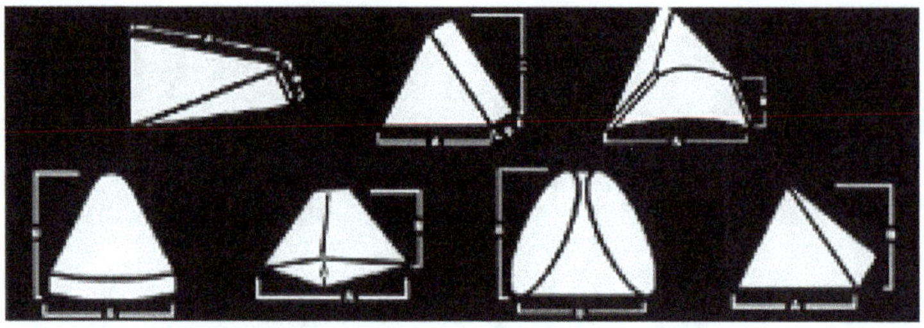

Plastic medias in most cases do not cut as fast as ceramic but do give lower R.M.S. finishes and eliminates the nicking or damaging of edges on softer metals.

Below are a number of advantages of plastic media:

- Light weight and soft bonding agents allow it to wear a burr off with out rolling the burr onto the part. Once a burr is rolled it's very difficult to remove.
- Certain plastic shapes, particularly a wedge, will sharpen as it wears allowing points to get into inside radious of parts. In large mass finishing systems with large parts, hard medias compressed between the machines sidewall and the rotating part can gouge and damage the part. Softer plastic medias eliminate this type of damage.
- Low RMS finishes.

Plastic media has many advantages. Your parts should be run in a lab prior to media selection to determine its future success.

Random Medias

Random medias are comprised of a variety of natural shapes and types of materials, one of the oldest recorded applications of random media used in mass finishing were by the Romans. They gathered certain types of rock, put them in pouches attached to their belts and polished their uniform ornaments as they marched.

Today, some popular random medias are microcrystalline aluminum oxide, silicon carbide and sintered ceramic. These medias are tough, chemically inert, economical, non-toxic and uniformally sized.

Random medias can cause more lodging problems than the more commonly used preformed medias.

Dry Process Media

The most common dry process media are wood pegs, corncobs and walnut shells. They are produced in a variety of shapes and sizes.

The dry process medias are most often impregnated with dry polishing powders or creams. The dry polishing powders are chrome rouge, iron oxide or calcide aluminum that are normally impregnated by the media manufacture. The polishing cream also contains polishing powder combined with other elements that are added to the dry process media by the user and recharged between part batches.

Dry medias are used in most mass finishing machines, including: vibratory, rotating barrels, centrifugal barrels, centrifugal disc and spindle machines.

The most common applications of dry finishing media is surface brightening (an as buffed finish) on final finish refinement. The dry media process can be a one step process, however, it's usually the last step of a two-step finishing process. Cleaning (especially with the creams) may also be required.

Steel Media

Steel media is a cold headed, heat treated, ground and finished carbon or stainless steel shape.

It is highly finished. Steel media is used in vibratory or barrel mass finishing to burnish, clean, improve compressive strength, dull edges, smooth finishes and reduces porosity on plated parts.

Steel media can last 10 years, runs very clean and eliminates lodging of media in parts. The media is available in a variety shapes and sizes.

The weight is approximately 300 lbs per cu ft. Specially built heavy duty machines are usually required to run steel because of it's weight.

Process times are relatively short and waste treatment is simplified because of nonsediment waste stream.

Soap Compound

Soap compounds are critical and in many cases have more importance in obtaining processes and finishes in mass finishing than the media itself.

Soap compounds accomplish many things:

- Lubricates the mass resulting in extended media life.
- Cushions the part eliminating part damage.
- Suspends the dirt, oils and media residues and flushes them from the system keeping the parts and media clean.
- Inhibits rust or corrosion of parts in the finishing process. Extended shelf life inhibiting should be done as a post dip or spray with water-soluble inhibitors.
- Burnishing (Brighting).
- Descaling.
- Accelerates media cut.

Soap compounds are identified in the following categories though many compounds will do multiple functions:

- Deburring compounds (General Purpose).
- Burnishing compounds (Brighting).
- Rust inhibitors (Post operation sprays or dips).
- Chemical accelerations (Metal oxidizes).
- Descalers.

Soap should be mixed with water and then flowed thru the system. The soap and water in many cases can pass straight to sewer and now more commonly recirculated back thru the system after being filtered.

The general soap concentration and water flows are 1 to 2 oz of soap per gallon of water and 1 to 2 gallons of water per cubic ft per hour.

A good starting point for the following processes are as follows:

Ceramic Media - 1 oz of soap per gallon of water and 1 gallon of water per cubic foot per hour.

Plastic Media - 1 oz of soap per gallon of water and 1 1/2 gallons of water per cubic foot per hour.

Dense Media - 1 oz of soap per gallon of water (small shapes) and 1/2 gallon of water per cubic foot.

You will need to increase these minimums as parts get dirtier or oilier.

Mass Finishing Work Load

The workload or amount of parts that can be processed in a given batch will largely determine the finishing cost of the part.

Factors that affect the workload are:

- Part size and configuration.
- Part weight.
- Alloy of material part is make of.
- Media to be used.

General comments about workload:

- Larger parts and or sharp cornered parts can be more susceptible to part on part damage.
- Heavy parts may require larger media to support and suspend the part within the mass.
- Smaller media (a dense mass) provides more protection, therefore allowing more parts to be run without damage.
- Parts that require very fine finish refinements or bright finishes cannot be loaded as heavy as a deburring application.

The workload is determined by volume and is expressed by media to part ratio.

Media to part ratio guidelines are as follows:

- Part on part processing using no media.

  0:1

- Very small stamping, forging, castings.

  1:1

- Very small machined parts.

  2:1

- Minimum amount required for self unloading machines.

  3:1

- Average part.

  4:1

- Softer non ferrous metals.

  5:1

- Very fine finishes and bright burnished parts.

  7:1

- Very delicate part.

  10:1

- Parts that cannot touch one another compartmentized.

Using media to parts ratio to calculate parts per cubic foot:

Once you've estimated the media to part ratio one can calculate the parts per cubic foot by multiplying part L" x W" x H" x media to part ratio (on ratio 4:1 use 5) % 1728 = parts per cubic foot.

Calculation of machine size required:

Example: use the following parameters.

- If you've estimated 12 parts per cubic foot.
- One hour total cycle (50 Min. process, 10 Min. unload).
- Daily production 1000 parts 1 shift, on 7 one-hour cycles. Divide 1000 by 7 = 143 parts per cycle.

- Machine size required 143 divided by 12 (Parts per cu. ft.) it will require a 12 cu. ft. machine to process the 1000 parts per day.

Deburring

Deburring is one of the most common applications of mass finishing. Below is general information on deburring applications:

- External burrs are easier removed than internal burrs.
- Burrs can be created too large to remove.
- Small parts with large burrs can be difficult requiring longer time cycles or higher energy machines.
- Burrs can be rolled by heavier medias therefore requiring lightweight medias to cut them instead of rolling them. Once a burr is rolled and flatted to the part it becomes impossible to remove by mass finishing.
- Stainless burrs will work harden themselves off the part instead of being worn off like other types of metals.
- Larger media, faster cutting and heavier media will remove larger burrs.
- Increased time cycles will assist in removing larger burrs.

Surface Finishing

Surface finishing in mass finishing refers to refining or smoothing a surface to lower the RMS finish. Pre plate finishes are often referred to in the industry as surface refinement.

Various medias will affect surfaces in different ways. Medias that are sharp cornered or harder than the work pieces can nick and damage parts.

Ceramic medias may do an excellent job of bringing an RMS from 64 to 16 but a secondary plastic media cycle may be required to bring the surface from 16 to 6. High-energy systems that flow media and or special medias will be required to bring surfaces in the 2 to 3 RMS range.

Time cycles can become extended in surface refinement ranging between 2 to 6 hours in vibratory or one hour in high-energy systems.

Burnishing

Burnishing in mass finishing is to brighten a work piece surface producing a luster finish.

Burnishing medias are primarily steel media, ceramic porcelain or dry medias. The process cycle times are usually one hour or less.

A bright surface will amplify scratches, nicks or imperfections; therefore, in many cases require a cut and smooth cycle prior to the burnishing cycle. The cut and smooth cycle can be fairly long, one to three hours in vibratory and is commonly done with a plastic media.

HONING

Honing is a process which is used to provide better surface finish and highly accurate work piece. It is a finishing or we can say super-finishing process used for finishing round holes by means of an abrasive material. This process uses a cylindrical mandrel dressed with an abrasive material like aluminium oxide, silicon carbide, diamond etc. This tool is called hone. This tool carried out a combined rotary and reciprocating motion to remove material at whole length of stationary work piece. The tool rotates at its own axis and reciprocates in the cylindrical hole to cover whole length of work piece. Mainly honing is considered as follow up process after drilling or boring. Sometimes it is used for finishing external surfaces.

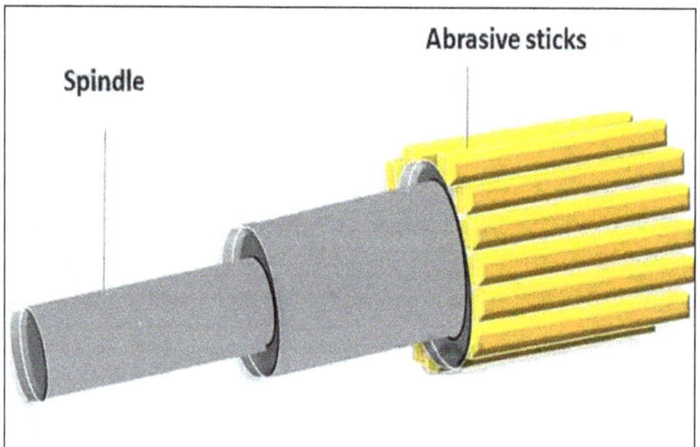

Hone

Honing can be considered as a machining process because it involves metal removal in form of microchips. The word "Honing" is developed by the word "Hone". Hone is used for a cylindrical tool dressed with some abrasive material. This abrasive material is used in the form of sticks which are mounted on the mandrel. However nowadays diamond and carbon boron nitrite grits are used for complete the operation in just one stroke. This process is used for achieve accurate dimension of cylindrical holes. It is best suited for finishing engine cylinder. Sometimes It is used for finishing gear where tool made with plastic impregnated with abrasive.

In honing process a well-dressed (with abrasive particles) cylindrical tool rotates at its own axis. The tool diameter is almost equal to the work piece internal diameter. This tool rotates at its own axis and also reciprocates which allows to remove material in form of microchips. It can achieve tolerance of about 0.25 – 1.5 micron. This cutting form a cross lay hatched pattern at work piece.

The accuracy and surface finish of hole is mainly depends on various process parameter which are describe below.

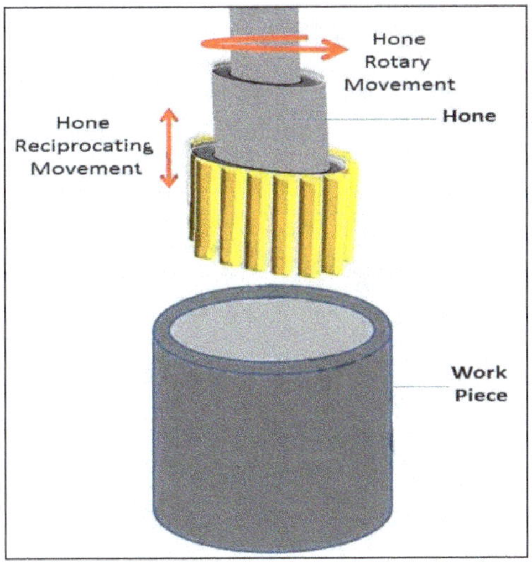

Process Parameters

RPM of Tool

If the tool speed increases, it increases metal removal rate and decreases surface roughness.

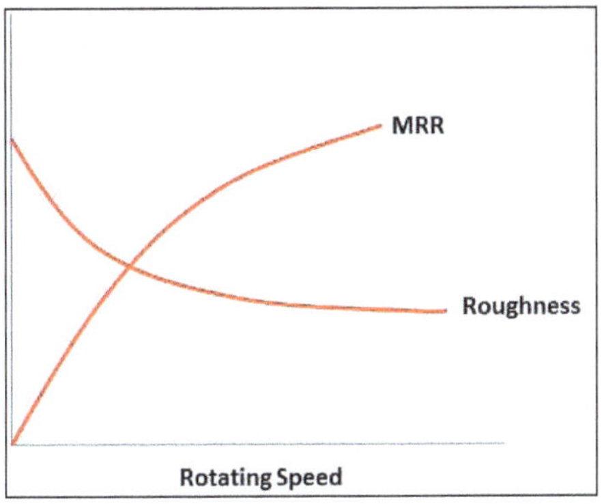

Honing Time

If we increase honing time it will decrease surface roughness at some extent and then starts increases it. It is so selected which gives minimum surface roughness with maximum metal removal rate.

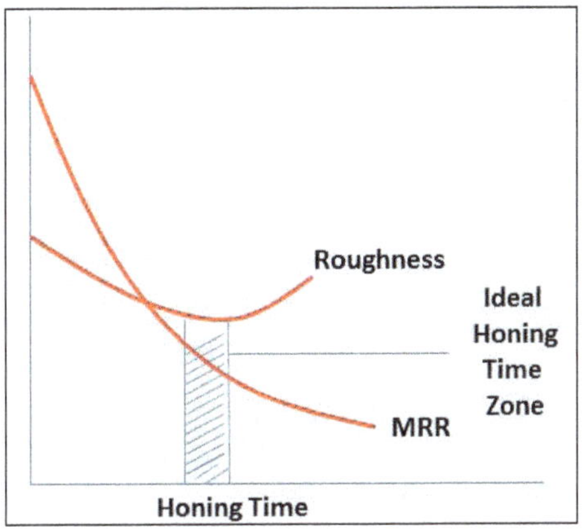

Honing Stick Pressure

The change in metal removal rate and surface roughness with the change in unit pressure is shown in the graph. The honing stick pressure is so selected as to get minimum roughness with highest metal removal rate.

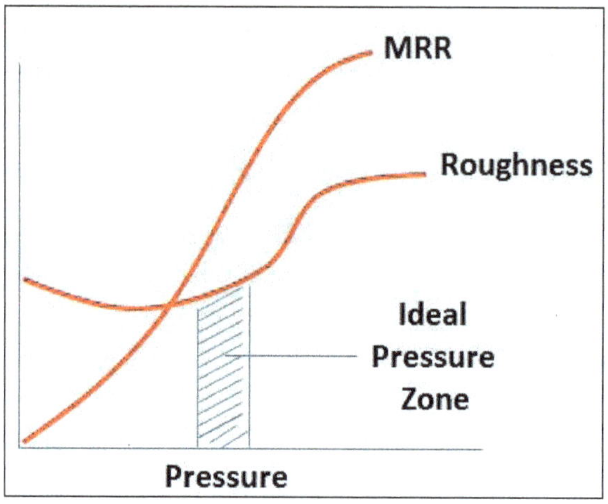

Stroke Length

The stroke length should be as enough which can cover the whole working length.

Types

Honing can be classified into two types.

Manual Honing

In this type of honing the hone rotate continuously and the work piece is move front and back manually. This process is rarely used now days.

Machine Honing

The honing process in which the hone performs combined rotary and reciprocating motion and there is less manual interaction called machine honing. According to the type of machine this process can be classified into two types:

- Vertical honing.
- Horizontal honing.

Advantages

- Less complex or low cost fixtures.
- It is highly accurate.
- It can be used for both long and short bores.

- It maintains original bore centerline.
- Any material can be finished regardless its hardness.

HOBBING

Hobbing is a machining process for gear cutting, cutting splines, and cutting sprockets on a hobbing machine, which is a special type of milling machine. The teeth or splines of the gear are progressively cut into the material (a flat, cylindrical piece of metal) by a series of cuts made by a cutting tool called a hob. Compared to other gear forming processes it is relatively inexpensive but still quite accurate, thus it is used for a broad range of parts and quantities.

It is the most widely used gear cutting process for creating spur and helical gears and more gears are cut by hobbing than any other process as it is relatively quick and inexpensive.

A type of skiving that is analogous to the hobbing of external gears can be applied to the cutting of internal gears, which are skived with a rotary cutter (rather than shaped or broached).

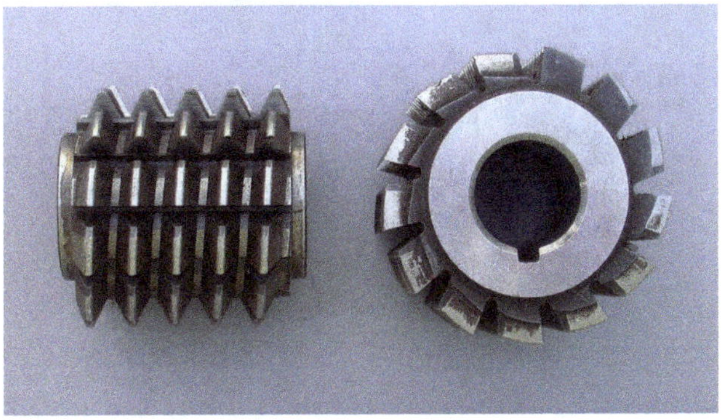

A hob — the cutter used for hobbing.

Process

Hobbing uses a hobbing machine with two skew spindles, one mounted with a blank workpiece and the other with the hob. The angle between the hob's spindle (axis) and the workpiece's spindle varies, depending on the type of product being produced. For example, if a spur gear is being produced, then the hob is angled equal to the lead angle of the hob; if a helical gear is being produced then the angle must be increased by the same amount as the helix angle of the helical gear. The two shafts are rotated at a proportional ratio, which determines the number of teeth on the blank; for example, for a single-threaded hob if the gear ratio is 40:1 the hob rotates 40 times to each turn of the blank, which produces 40 teeth in the blank. If the hob has multiple threads the speed ratio must be multiplied by the number of threads on the hob. The hob is then fed up into the workpiece until the correct tooth depth is obtained. Finally the hob is fed through the workpiece parallel to the blank's axis of rotation.

Often multiple blanks are stacked, then cut in one operation.

For very large gears the blank can be gashed to the rough shape first to make hobbing easier.

Equipment

A horizontal hobbing machine.

A horizontal hobbing machine.

Hobbing machines, also known as *hobbers*, are fully automated machines that come in many sizes, because they need to be able to produce anything from tiny instrument gears up to 10 ft (3.0 m) diameter marine gears. Each gear hobbing machine typically consists of a chuck and tailstock, to hold the workpiece or a spindle, a spindle on which the hob is mounted, and a drive motor.

For a tooth profile which is a theoretical involute, the fundamental rack is straight-sided, with sides inclined at the pressure angle of the tooth form, with flat top and bottom. The necessary addendum correction to allow the use of small-numbered pinions can either be obtained by suitable modification of this rack to a cycloidal form at the tips, or by hobbing at other than the theoretical pitch circle diameter. Since the gear ratio between hob and blank is fixed, the resulting gear will have the correct pitch on the pitch circle, but the tooth thickness will not be equal to the space width.

Hobbing machines are characterised by the largest module or pitch diameter it can generate. For example, a 10 in (250 mm) capacity machine can generate gears with a 10 in pitch diameter and usually a maximum of a 10 in face width. Most hobbing machines are vertical hobbers, which means the blank is mounted vertically. Horizontal hobbing machines are usually used for cutting longer workpieces; i.e. cutting splines on the end of a shaft.

Hob

A gear hob in a hobbing machine with a finished gear.

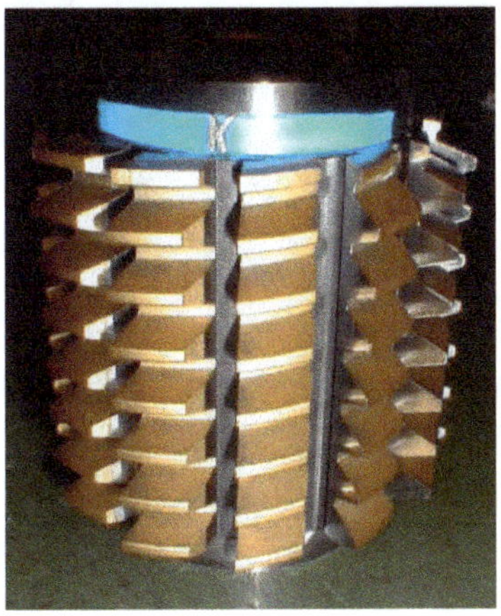

The *hob* is a cutting tool used to cut the teeth into the workpiece. It is cylindrical in shape with helical cutting teeth. These teeth have grooves that run the length of the hob, which aid in cutting and chip removal. There are also special hobs designed for special gears such as the spline and sprocket gears.

The cross-sectional shape of the hob teeth are almost the same shape as teeth of a rack gear that would be used with the finished product. There are slight changes to the shape for generating purposes, such as extending the hob's tooth length to create a clearance in the gear's roots. Each hob tooth is relieved on the back side to reduce friction.

Most hobs are single-thread hobs, but double-, and triple-thread hobs increase production rates. The downside is that they are not as accurate as single-thread hobs. Depending on type of gear teeth to be cut, there are custom made hobs and general purpose hobs. Custom made hobs are different from other hobs as they are suited to make gears with modified tooth profile. The tooth profile is modified to add strength and reduce size and gear noise.

This list outlines types of hobs:

- Roller chain sprocket hobs.
- Worm wheel hobs.
- Spline hobs.

- Chamfer hobs.
- Spur and helical gear hobs.
- Straight side spline hobs.
- Involute spline hobs.
- Serration hobs.
- Semitopping gear hobs.

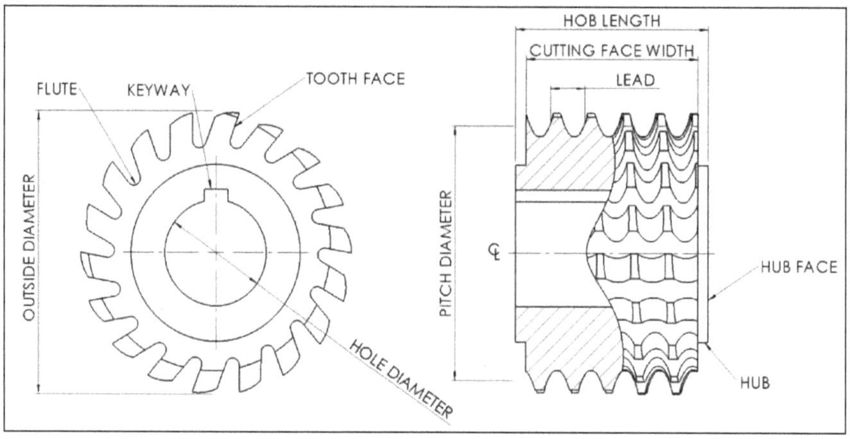

Uses

Hobbing is used to make the following types of finished gears:

- Cycloid gears.
- Helical gears.
- Involute gears.
- Ratchets.
- Splines.
- Sprockets.
- Spur gears.
- Worm gears.

Hobbing is used to produce most throated worm wheels, but certain tooth profiles cannot be hobbed. If any portion of the hob profile is perpendicular to the axis then it will have no cutting clearance generated by the usual backing off process, and it will not cut well.

Cycloidal forms

For cycloidal gears (as used in BS978-2 Specification for fine pitch gears) and cycloidal-type gears each module, ratio and number of teeth in the pinion requires a different hobbing cutter, so the technique is only suitable for large volume production.

To circumvent this problem a special war-time emergency circular arc gear standard was produced giving a series of close-to-cycloidal forms which could be cut with a single hob for each module for eight teeth and upwards to economize on cutter manufacturing resources. A variant on this is still included in BS978-2a (Gears for instruments and clockwork mechanisms. Cycloidal type gears. Double circular arc type gears).

Tolerances of concentricity of the hob limit the lower modules which can be cut practically by hobbing to about 0.5 module.

ELECTROCHEMICAL MACHINING

Electrochemical Machining is a non-traditional machining process in which metal removed by electrochemical dissolution.

Principle of Electrochemical Machining

Electrochemical machining works on the Faraday law of electrolysis which state that if two electrode are placed in a container which is filled with a conductive liquid or electrolyte and high ampere DC voltage applied across them, metal can be depleted form the anode (Positive terminal) and plated on the cathode (Negative terminal). This is the basic principle of electrochemical machining. In this machining process, tool is connected with the negative terminal of battery (work as cathode) and work-piece is connected with the positive terminal of battery (work as anode). They both are placed in a electrolyte solution with a small distance. When the DC current supplied to the electrode, metal removed from work-piece. This is basic fundamental of electrochemical machining.

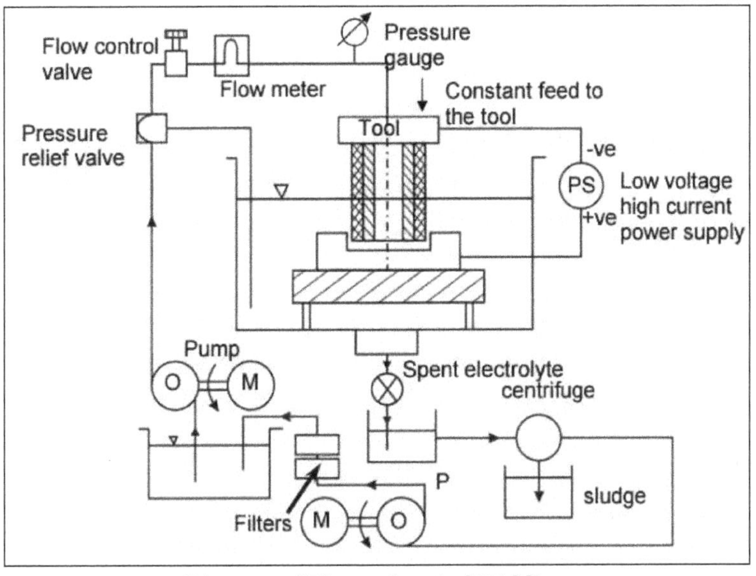

Diagram of Electrochemical Welding.

Equipment

Power Supply

In electrochemical machining process, a high value of direct current around 40000A and low value of potential difference around 10-25V is desirable. The electrodes are place at a inter electro gap witch is desirable for machining. If the inter electro gap not too small witch can generate arc or not too high witch is not suitable for machining. It is about 1mm. This high values DC current is form by convert three phase AC current into DC current by using Silicon Controlled Rectifier.

Electrolyte Supply and Cleaning System

It consist piping system, storage tank, pump, control valve, pressure gauge, heating or cooling coil etc. in the electrochemical process, the metal removed from work-piece form sludge which should be remove form electrolyte. This system control the flow and cleaning of electrolyte solution into the container. Piping system is made of SS steel, Glass fiber reinforced plastic, plastic lined MS or similar other anti-corrosive material. The tank capacity is about 500 gallon for per 10000A of current.

Tool and Tool Feed System

Tool is made by an anti-corrosive material because it has to withstand in corrosive environment for long time. It should also have high thermal conductivity and easily machinable. The dimension accuracy and surface finish of work piece is directly depends on tool dimension. Those part of the tool, which is not required for machining, should be properly insulated because lack of insulation tends to unwanted machining which give dimensional inaccuracy.

Work Piece and Work Holding System

In this process, work piece should be well electric conductive. Only electric conductive material can be machined by this method. Work piece take as anode in this process. The work holding devices should have non-conductive property.

Working of Electrochemical Machining

Electrochemical machining works inverse as electroplating process. Metal is removed form anode into electrolyte and convert into slag form by reacting opposite ions available in electrolyte. This process works as follow.

- In ECM, the electrolyte is so chosen that there is no plating on tool and shape of tool remain unchanged. Generally NaCl into water takes as electrolyte.
- The tool is connected to negative terminal and work is connected to positive terminal.
- When the current passes through electrode, reaction occur at anode or workpiece and at the cathode or tool. To understand proper working let's take an example or machining low carbon steel.
- Due to potential difference ionic dissociation take place in electrolyte.

$NaCl \leftrightarrow Na^+ + Cl^-$

$H_2O \leftrightarrow H^+ + OH^-$

- When the potential difference applied between the work piece and tool, positive ions move towards the tool and negative ions towards the work piece.
- Thus the hydrogen ion moves towards tool. As the hydrogen reaches to the tool, it takes some electron from it and converts into gas form. This gas goes into environment.
- When the hydrogen ions take electron from tool, it creates lack of electron in mixture. To compensate it, a ferrous ions created at the work piece (anode) which gives equal amount of electron in mixture.

  $2H^+ + 2e^- = H_2 \uparrow$ at Cathode

  $Fe = Fe^{++} + 2e^-$ at Anode

- These Ferrous ions react with opposite chlorine ions or hydroxyl ions and get precipitate in form of sludge.

  Iron (Fe) $\leftrightarrow Fe^{++} + 2e^-$

  $Fe^{++} + 2Cl^- \leftrightarrow FeCl_2$

  $Fe^{++} + 2(OH)^- \leftrightarrow Fe\,(OH)$

  $Fecl_2 + 2(OH)^- \leftrightarrow Fe\,(OH)_2 + 2Cl$

- This will give ferrous or iron into electrolyte and complete the machining process. This machining process gives higher surface finish because machining is done atom by atom.

Application

- ECM is used to machining disk or turbine rotor blade.
- It can be used for slotting very thin walled collets.
- ECM can be used to generate internal profile of internal cam.
- Production of satellite rings and connecting rod, machining of gears and long profile etc.

Advantages

- It can machine very complicated surface.
- A single tool can be used to machining large number of work-piece. Theoretically no tool wear occur.
- Machining of metal is independent on strength and hardness of tool.
- ECM gives very high surface finish.

Disadvantages

- High initial cost of machine.
- Design and tooling system is complex.
- Fatigue property of machined surface may reduce.
- Nonconductive material cannot be machined.
- Blind hole cannot be machined form ECM.
- Space and floor area requirement is high compare to conventional machining.

References

- Paris, Henri (2005). "Modelling the Vibratory Drilling Process to Foresee Cutting Parameters". Cirp Annals. 54: 367–370. Doi:10.1016/S0007-8506(07)60124-3
- Types-machining, custom-manufacturing-fabricating: thomasnet.com, Retrieved 21 August, 2019
- Mattson, Mike (2009), CNC Programming: Principles and Applications (2nd ed.), Cengage Learning, p. 233, ISBN 978-1-4180-6099-2
- What-is-honing-its-process-parameter-types-and-advantages: mech4study.com, Retrieved 22 January, 2019
- Koepfer, Chris (2010-01-22). "Hard Turning as an Alternative to Grinding". Production Machining. Retrieved 2010-03-04
- Electrochemical-machining-principle-working-equipment-application-advantages-and-disadvantages: mech-4study.com, Retrieved 23 February, 2019

PERMISSIONS

All chapters in this book are published with permission under the Creative Commons Attribution Share Alike License or equivalent. Every chapter published in this book has been scrutinized by our experts. Their significance has been extensively debated. The topics covered herein carry significant information for a comprehensive understanding. They may even be implemented as practical applications or may be referred to as a beginning point for further studies.

We would like to thank the editorial team for lending their expertise to make the book truly unique. They have played a crucial role in the development of this book. Without their invaluable contributions this book wouldn't have been possible. They have made vital efforts to compile up to date information on the varied aspects of this subject to make this book a valuable addition to the collection of many professionals and students.

This book was conceptualized with the vision of imparting up-to-date and integrated information in this field. To ensure the same, a matchless editorial board was set up. Every individual on the board went through rigorous rounds of assessment to prove their worth. After which they invested a large part of their time researching and compiling the most relevant data for our readers.

The editorial board has been involved in producing this book since its inception. They have spent rigorous hours researching and exploring the diverse topics which have resulted in the successful publishing of this book. They have passed on their knowledge of decades through this book. To expedite this challenging task, the publisher supported the team at every step. A small team of assistant editors was also appointed to further simplify the editing procedure and attain best results for the readers.

Apart from the editorial board, the designing team has also invested a significant amount of their time in understanding the subject and creating the most relevant covers. They scrutinized every image to scout for the most suitable representation of the subject and create an appropriate cover for the book.

The publishing team has been an ardent support to the editorial, designing and production team. Their endless efforts to recruit the best for this project, has resulted in the accomplishment of this book. They are a veteran in the field of academics and their pool of knowledge is as vast as their experience in printing. Their expertise and guidance has proved useful at every step. Their uncompromising quality standards have made this book an exceptional effort. Their encouragement from time to time has been an inspiration for everyone.

The publisher and the editorial board hope that this book will prove to be a valuable piece of knowledge for students, practitioners and scholars across the globe.

INDEX

R

S

T

V

Z

www.ingramcontent.com/pod-product-compliance
Lightning Source LLC
LaVergne TN
LVHW080922210926
848485LV00004B/711
* 9 7 8 1 6 4 7 2 8 3 3 3 9 *